Electronic Post-Production and Videotape Editing

Electronic Post-Production and Videotape Editing

Arthur Schneider

Focal Press
Boston London

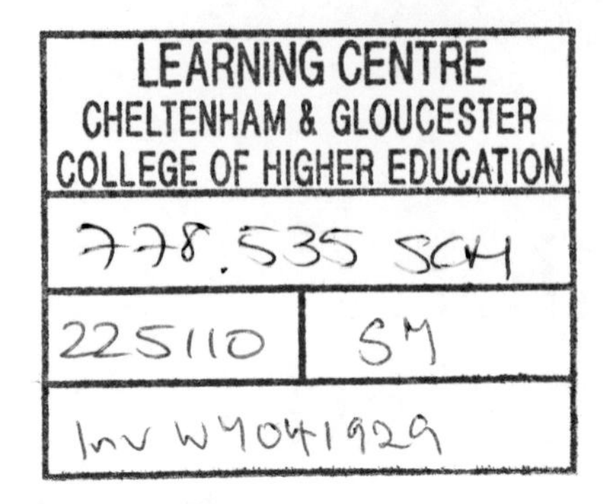

Focal Press is an imprint of Butterworth Publishers.

Library of Congress Cataloging-in-Publication Data

Schneider, Arthur, 1930-
Electronic post-production and videotape editing / Arthur Schneider.
p. cm.
Bibliography: p.
Includes index.
ISBN 0-240-51799-7
1. Video tapes—Editing. I. Title.
TR899.S36 1989
778.59'92—dc19 88-24162
CIP

British Library Cataloguing in Publication Data

Schneider, Arthur
Electronic post-production and videotape editing.
1. Videorecordings. Editing. Applications of electronic equipment
I. Title
778.5'35
ISBN 0-240-51799-7

Butterworth Publishers
80 Montvale Avenue
Stoneham, MA 02180

10 9 8 7 6 5 4 3 2 1

Printed in the United States of America

This book is dedicated to my family and the students I have taught over the years. Both my children are working in the television industry. My son, Robert, is a videotape editor in his own right, and my daughter, Lori, is a videotape operator. Both are respected within the industry as highly talented individuals. I also wish to thank my wife, Dee, for her help and support during the years it has taken to get this book into publication.

I have been gratified to see a number of my students get jobs in the field of post-production. I encourage those of you determined to get into this fascinating industry of editing and post-production to keep at it and hope this book will assist you in reaching that goal.

Contents

Foreword

Art Schneider and I first met more than 30 years ago when he was a young film editor working at NBC in Hollywood. He became a member of our team soon after and has edited nearly 50 of my television shows on film and videotape over the years, including several of my overseas Christmas specials.

Art has shown great skill as an editor and many times has fixed my mistakes by skillfully editing out my not so funny jokes. For that effort, in 1965 during the production of one of my comedy specials, I presented Art with a plaque that reads, "The Bob Hope Show Proudly Presents the Crossed Scissors Award With Seeing Eye Cluster to Art Schneider For Jump Cutting Above and Beyond the Call of Duty."

Our close working relationship over the years has shown me that Art is a talented craftsman always eager to help others. His son, Robert, has moved into his dad's shoes and now edits my specials, with more than 30 shows to his credit. His nickname is "Jump Cut Junior." Next the grandchildren!

I know this book will be helpful to others in their pursuit of a career in editing.

Bob Hope

Preface

Videotape editing is one of the least understood and seemingly most complex of all the technical crafts in the television industry. Since the advent of the first electronic videotape editing system in 1962, the demand for people and equipment has continued to grow, as more and more people in the industry have turned to videotape as a fast, cost-efficient communications medium.

There is something frightening and magical about the term *videotape editing*. Many people just getting into the field are intimidated by the jargon they find in most technical books, but in this book, I approach the post-production process in a simple, direct manner. By unveiling the so-called mysteries of videotape editing, I hope to give you an opportunity to explore the many aspects of editing and post-production and, in turn, to acquaint you with the possibility of pursuing a creative and fulfilling career in the field.

You might have found there is little current printed material on post-production and videotape editing. Part of the reason is that the television industry is changing so rapidly that it is difficult to keep up with these changes. Part of this book is devoted to the history of videotape editing in television, while the rest of the book describes editing techniques that will help you develop into an efficient videotape editor. The last chapter speculates on things to come.

My suggestions apply to any style of editing, whether it is for educational or industrial use, syndication, or commercial broadcast. The basic functions of the editing process are outlined so that it is easy to follow the sequence of building an edited work tape.

This book is not designed to show you how to edit. That task is better left to a formal hands-on training program where you can experiment and work with editing equipment. The editing systems available range from simple cuts-only systems to those that come complete with all the bells and whistles. My intention is to acquaint you with the basic features of many of the more popular systems and to describe techniques that will help you during the editing process. Although I strive to be nontechnical, I do introduce some technical terms and procedures you should know if you intend to be an efficient and productive videotape editor. These are all defined in the glossary at the end of the book.

Having been involved in motion picture film and videotape editing for more than 37 years, I have been fortunate to see the exciting advancements in this fascinating industry and to have been a pioneer in the development of many of the editing tools we use today. I have no secrets about editing, but I hope that the knowledge I have gained about videotape editing will help you to become a successful videotape editor.

Acknowledgments

This book was made possible through the efforts of numerous individuals and corporations involved in the broadcast and videotape industries. Because of obvious space limitations, I cannot list all the individuals and companies that helped me, but those I do not mention should know how much I appreciate their efforts.

My sincere thanks go to Ampex Corp.; CMX Corp.; Convergence Corp.; Ediflex; EECO, Inc.; The Grass Valley Group, Inc.; MCI/Quantel; Fernseh, Inc.; Montage Computer Corporation; Sony Broadcast Products Co.; and Tektronix. I would also like to thank the following individuals for their encouragement and input: Dave Bargen, Jack Calaway, Sil Caranchini, Rome Chelsi, Emory Cohen, Craig Curtis, Christin Hardman, Bob Ringer, and Jan Yarbrough.

Introduction

When Art Schneider first came to the hallowed halls of the National Broadcasting Company in the early 1950s, videotape was just beginning to make an impact on the way television was produced. I doubt we could have imagined that it would someday be the standard that it is now. Art was part of a small group that developed the double system editing method, the first of the off-line videotape editing concepts. As he details in the book, videotape editing at this point was a much closer relation of film editing than it is today. Being new, videotape was regarded by most with a mixture of interest and suspicion. Thanks to the efforts of the people involved in developing it, however, videotape became much easier to use.

Our little group of specialists, which comprised both technical and editorial talent, generated an enthusiasm that was self-sustaining. Those of us involved realized, even if many in the industry did not, the creative potential for the videotape editor. Not long after the birth of the double system editing method, almost all the top talent from many networks and studios edited with this process at NBC.

Through the golden years of the 1950s, 1960s, and 1970s, videotape became more and more widely used for television production and broadcasting. Slowly, the role of the videotape editor expanded from simple technical functions to a form of artistic expression. Electronic editing and the development of time code produced new possibilities as well as new challenges.

As technology has grown ever more dynamic and complex and the tasks of the editor have expanded, the need for education has become crucial to new editors. As Art explains so well, the successful editor of today requires creative, personal, and business abilities far beyond the necessary technical skills. This book incorporates all of these aspects and is thus a building block to help people understand the need for and use of the videotape editing process.

I suggest that this book be studied not only by present and future editors, but also by producers, directors, and production managers who must understand the post-production process to do an effective job. This book has been written by an expert in the field, a man with some 37 years of experience who has been involved in hundreds of television editing assignments. He provides a distillation of this tremendous knowledge in this book, and I believe you will both enjoy and learn from his efforts.

Craig Curtis
Director of Recording and Post-Production
NBC Burbank

Communications, Television, and Tape Editing

1

Since prehistoric times, people have tried to express themselves. Paintings on cave walls and stone carvings were two of the ways early people expressed their desire to communicate and to preserve their thoughts and ideas for other people.

As time went by, people's ability to communicate gradually improved. One big step in the development of mass communication was the invention of the camera. During the more than 220 years that people have used the camera to record images, photography has changed dramatically.

When photography first came on the scene in 1772, photographs were made using a mixture of silver nitrate and chalk. Over the years, technical improvements led to the process of recording images on metal, glass, paper, and film bases.

But photography was not the only development that changed the way people recorded the world around them. Looking back in history, mass communication might have started in 1877 when Emile Berliner invented the microphone. About the same time, Thomas A. Edison introduced the phonograph. A few years later, in 1893, Edison invented a device that made still photographs come alive. Edison's invention of the motion picture allowed us to record thousands of images on a narrow strip of celluloid and play them back on a screen.

In 1906 Lee DeForest invented the triode, a three-element vacuum tube that led to the invention of the radio. The radio brought news, special events, entertainment, and education into millions of homes. But people wanted more: They wanted to be able to see the things they were hearing on the radio.

In 1925 inventors such as J.L. Baird of England and C.F. Jenkins of the United States, working independently of one another, came up with the same invention—television. By today's standards, their efforts were crude, but they marked the beginning of a revolutionary communications medium.

Before 1927 both motion pictures and the first attempts at television had no sound. Any sound accompanying motion pictures had come from live performers, mostly piano players. In 1927 Warner Brothers produced a picture called *The Jazz Singer,* starring Al Jolson. It was the first commercially accepted motion picture to use sound synchronized with the picture (which we now call lip sync) and was considered to be the first talking motion picture. The next year, the Technicolor Corporation produced the first color motion pictures.

During the 1930s, experimental television broadcasting was conducted in the United States, but it wasn't until July 1, 1941, that the first commercial television broadcast was transmitted by station WNBT in New York. At about the same time, CBS and RCA were perfecting color television and trying to get the Federal Communications Commission (FCC) to accept their respective systems as the standard for color television production in the United States. RCA finally won out, and in 1955 the first color television pictures were transmitted halfway across the United States from RCA in New York to the 3M Company in St. Paul, Minnesota.

ZONE DELAY RECORDINGS

In the early 1950s, coaxial cable was the only way of transmitting coast-to-coast television programs. But coaxial cable could transmit only live programs, which posed a problem on both coasts. If a program was being aired live in New York at 8:00 p.m., the people on the West Coast would see the program three hours earlier, at 5:00 p.m.

To solve this problem, the networks devised a plan to photograph television programs on motion picture film as they were being broadcast live in the East. The film was rushed to the developing lab, processed, and returned to the network within three hours so that it could be put on a projector and shown to the West Coast audience at the same airtime as it was seen on the East Coast.

This was known as the *kinescope,* or *kine,* process. The term was derived from the name of the picture tube photographed by a special motion picture camera to generate a black-and-white motion picture film of the live television broadcast. (See Figure 1.1) Other terms for this process are electronic film recording (EFR), television recording (TVR), and video-to-film recording (VFR).

Briefly, here is how the process worked. A program aired live in New York at 8:00 p.m. was recorded on 35mm and 16mm black-and-white kinescope film for broadcast three hours later on the West Coast. If the program was longer than 30 minutes, a second pair of 35mm and 16mm cameras was started about 2 minutes before the first cameras ran out of film. Each 30-minute roll of film really contained about 33 minutes, so the changeover process allowed the engineers to record adequate overlaps.

As soon as the first half hour ran out, the exposed film was brought to the darkroom and unloaded. A messenger stood outside the darkroom door to rush it to the processing lab, which was about 15 minutes away from the studio. Two hours later, at about 7:00 p.m. (West Coast time), the messenger returned with the 35mm film under his arm. Film editors were standing by to splice identifying leaders onto the film and perform other functions needed to get the film ready for broadcast.

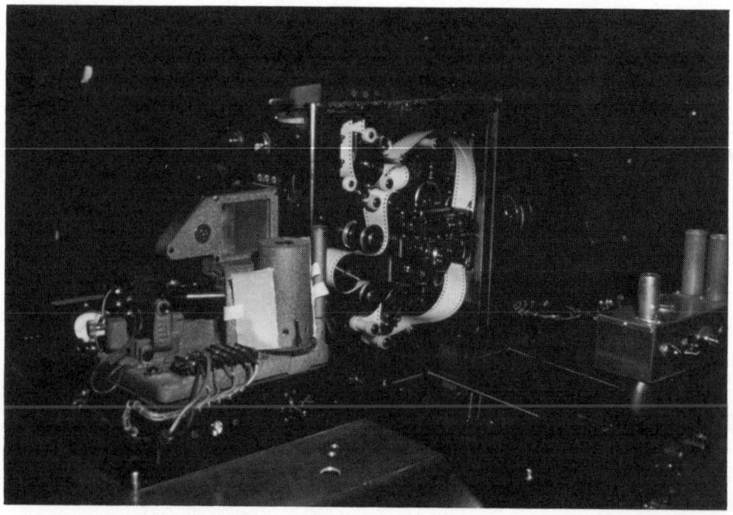

The 35mm film was exposed in order to generate a negative image that was then electronically reversed to produce a positive image on a television screen. The two reasons for this were (1) time constraints did not allow for making prints from the negative and (2) after broadcast, the negative could be used for printing. The 35mm negative film was sometimes used to make distribution prints for limited distribution by optically reducing the 35mm negative to a 16mm print format.

FIGURE 1.1
An example of a 35mm kinescope recording camera used for three-hour zone delays.

In addition to the 35mm negative, two types of 16mm kinescope recordings were made. One was called a *direct positive* print, which contained a positive picture and sound track suitable for direct projection onto a movie screen. Only one copy was made because its primary function was as a backup for the 35mm negative in the event of technical difficulties while on the air. The other, more cost effective process, utilized 16mm kinescope recordings for large syndication orders by recording a separate picture negative and an optical sound track negative at the time of the live broadcast to get optimum picture and sound quality. These two negatives were printed in a conventional manner at a film laboratory to produce 16mm composite picture and sound release prints for distribution to television stations on a two-week delayed basis throughout the United States and Canada. Not all programs were syndicated on a delayed basis; only those specifically ordered for syndication were made in this manner.

The 16mm backup copy was processed in-house and was ready to be sent to the projection room well before the 35mm film had returned from the lab. For programs longer than 30 minutes, standard reel changeover marks were added to the film so the projectionist would know when to switch from one reel to another.

Even though this process was pretty well defined, there were times in my own experience when we came close to not having the film ready on time. This was usually because of film breakage or other technical problems.

One of the worst situations I experienced occurred when the film broke in the developing machine and did not arrive from the lab until 15 minutes before airtime. We had to splice the film and tell the telecine operator where to switch to the 16mm direct positive protection copy until the damaged area had passed.

We delivered the film to the projectionist about one minute before airtime. The operator grabbed the film and threaded it in the gate of the projector. Just as he was starting to thread the film onto the take-up reel, the projector started. The program went on the air on time, but the operator was unable to get the film onto the take-up reel. Since the crew was not anxious to have 3,000 feet of motion picture film all over the floor, we formed a brigade to feed the film to a rewind bench where someone was slowly winding it onto an empty reel.

COLOR ZONE DELAYED RECORDINGS

Kinescope recording was destined to be replaced by higher-quality storage media. Although the 35mm version of the kinescope recording eventually attained the best quality the equipment could provide, it could not generate color.

By the time coaxial cable was replaced by a more sophisticated microwave link in the late 1950s, the public was clamoring for color. Although live broadcasts were being shown in color, zone delay recordings proved to be a serious problem for broadcast engineers.

Although attempts to make kinescope recordings in color were made, the long processing time for color film, along with its high cost, prevented it from being used for three-hour zone delay broadcasts. In 1952, NBC developed a system that could record color signals on special 35mm black-and-white film that had thousands of tiny lenses, or lenticules, embossed on the base, or celluloid, side of each frame. This was known as lenticular film (Figure 1.2).

To record the color signal, it first had to be broken down into the three primary colors—red, green, and blue. Each color signal was sent to individual black-and-white picture tubes photographed through respective red, green, and blue filters before being exposed through the lenticules and onto the film.

As each frame was being recorded, the lenticules broke down the color signal into minute areas of various shades of black, white, and gray that represented the color in the frame. Each frame was electronically exposed as a positive image, and after developing, the film was placed on a television projector with special red, green, and blue filters in front of the lens. This recombined the red, green, and blue elements of the picture. A color television camera photographed the resulting image and reproduced a full-color picture on the air. This process was used to broadcast color programs for several years.

FIGURE 1.2
A lenticular film frame. Special black-and-white film photographed through special color filters reproduces the image in color when sent through a telecine projector equipped with special color filters.

DEVELOPMENT OF VIDEOTAPE

The Ampex Corporation demonstrated the first broadcast black-and-white videotape recorder (VTR), the VR-1000 (Figure 1.3), in February 1956. Although earlier efforts had resulted in several other VTRs, Charles Ginsberg, Ray Dolby, Charles Anderson, and others at Ampex developed the first practical unit.

Ampex was bombarded with orders for videotape recorders at the National Association of Broadcasters (NAB) show in April. Broadcasters felt that this invention would enable them to record television pictures and sound on a single strip of two-inch-wide magnetic tape and, after rewinding, quickly play back what was just recorded. The videotape also was erasable and could be used over and over again.

Prior to the show, a survey had shown a five-year potential market of 26 machines. Amazingly, Ampex took orders for more than 75 re-

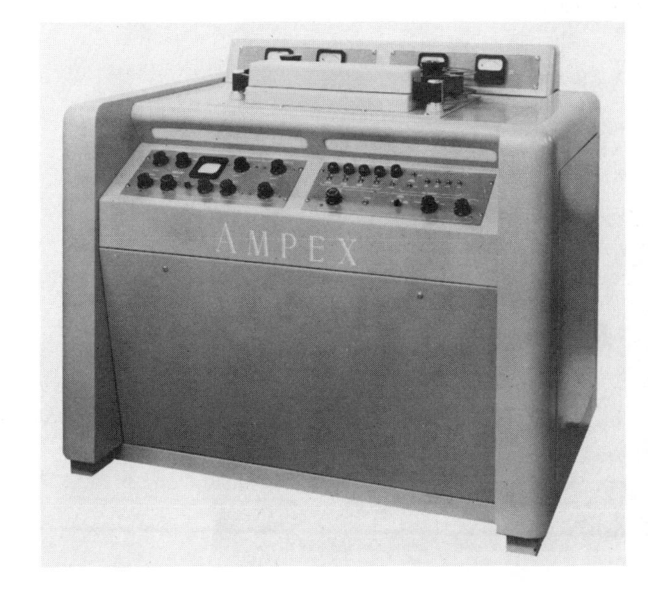

FIGURE 1.3
The first videotape recorder, the Ampex VR-1000. (Courtesy Ampex Corporation)

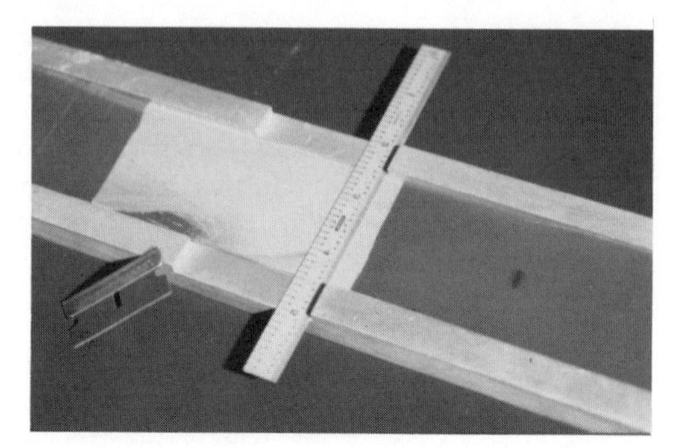

FIGURE 1.4 (above)
The RCA TRT-1 color videotape recorder.

FIGURE 1.5 (above right)
The first two-inch videotape splicer.

corders during that show alone. Within weeks of receiving their VTRs, the three television networks virtually stopped the kinescope recording of television programs on a zone delay basis.

In 1957, right on the heels of Ampex's videotape recorder, RCA announced the first color VTR, the TRT-1 (Figure 1.4). Ampex quickly followed with its version of color videotape, and the two companies agreed to share patents to make color videotape commercially available. By 1958 many prime-time television programs were being recorded in color.

Despite its advantages, this low-band color recording process left a lot to be desired. It did not provide good resolution, so the quality of subsequent copies declined rather quickly. A second-generation copy was marginally usable, but by the third generation pictures were snowy and had poor color rendition. These second- and third-generation copies were used only in the event of a technical failure of the on-the-air playback VTR.

In 1958, while employed by NBC in Burbank, California, I was involved in the development of videotape editing. In those days splicing together two-inch videotape required perfect eyesight, a sharp razor blade, and a lot of guts.

The first splicers (Figure 1.5) were made out of an aluminum block channeled out to hold the videotape. A solution of extremely fine iron particles known as carbonyl iron powder was mixed in a solution of Freon TF, a solvent used to clean the videotape machines. Using a tiny brush, this mixture was brushed over the magnetic oxide coating on the videotape (Figure 1.6). As the solvent evaporated, a pattern of the electronic signal was clearly visible on the tape.

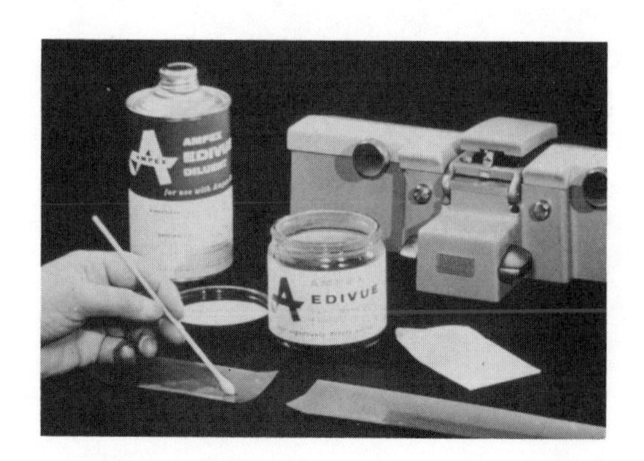

FIGURE 1.6 (bottom left)
Applying carbonyl iron solution on two-inch videotape.

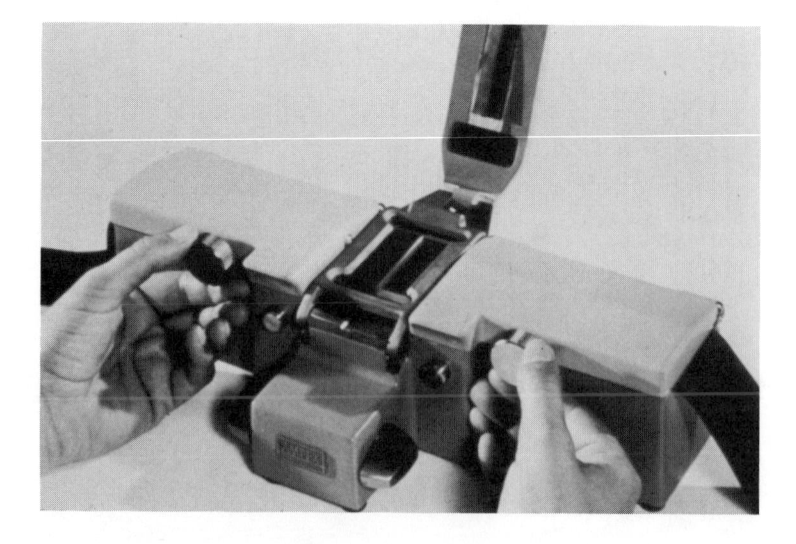

FIGURE 1.7 (bottom right)
Positioning the videotape in a splicer.

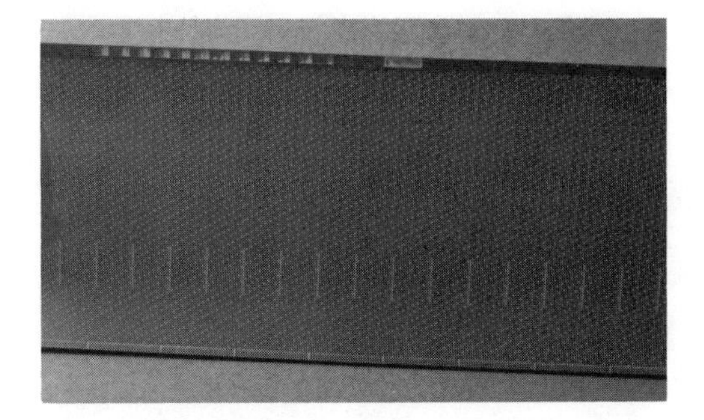

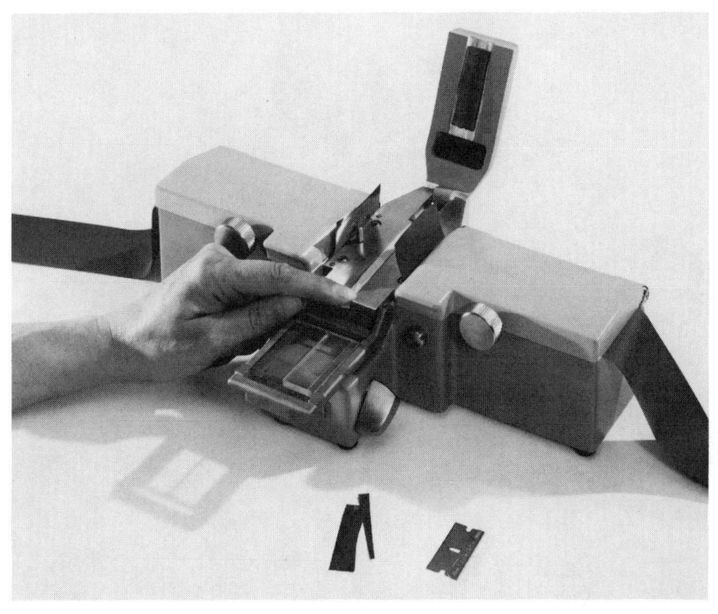

The editor would then locate the edit pulse, a much larger signal recorded on top of the control track that identified the frame line. He or she would position the edit pulse so that a steel ruler could be lined up with it. Pressing down firmly on the ruler, the editor would carefully and swiftly draw the edge of a razor blade across the width of the tape. (See Figure 1.7.)

Remember that the editor was working with extremely small tolerances and using the naked eye. For example, the dimensions of a complete television frame on two-inch videotape were about one and three quarters inches wide and half an inch long. Figure 1.8 shows the layout of the recorded signal on a piece of two-inch videotape.

Each hour of two-inch videotape was about 4,800 feet long. The video head recording or playing back the video images rotated at a rate of 14,400 revolutions per minute (rpm), or 240 revolutions per second. The linear speed of the tape was 15 inches per second. That translates to 30 television picture frames per second, or 60 fields per second, since each frame was made up of two fields.

To define a frame even further, its half-inch length was made up of 32 individual video tracks representing both television fields. These tracks were nearly perpendicular to the length of the tape. You could not cut in the middle of a television frame any more than you would try to make a splice in the middle of a film frame. Each of these 32 video tracks was about ten thousandths of an inch wide, and each track was separated by a blank guard band that was five thousandths of an inch wide. The guard band protected against interference from adjacent video tracks. This guard band was the only place two-inch tape could be spliced since it was an area of no recording and the video head did not pass over it.

On the bottom edge of the tape was the control track, which was similar to the sprocket holes on film. There were two audio channels on the tape, one for program and a secondary track used for time code. In the days before time code, the secondary audio channel was used to apply one frame audio tones, which triggered the beginning and end of an electronic edit using a system called Editec developed by Ampex.

FIGURE 1.8 (above left) Developed signal pattern on two-inch videotape.

FIGURE 1.9 (above right) Applying splicing tape to the back of the videotape. (Courtesy Ampex Corporation)

FIGURE 1.10 Splicing tape applied to base side of videotape. (Courtesy Ampex Corporation)

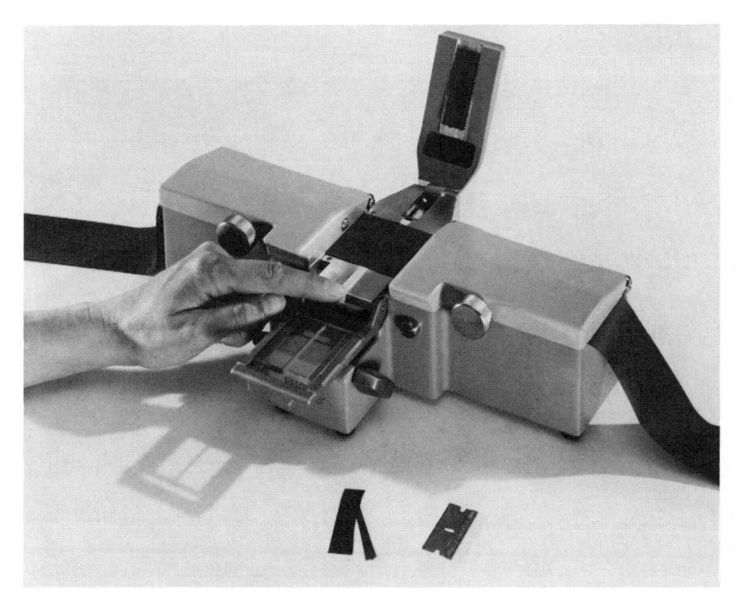

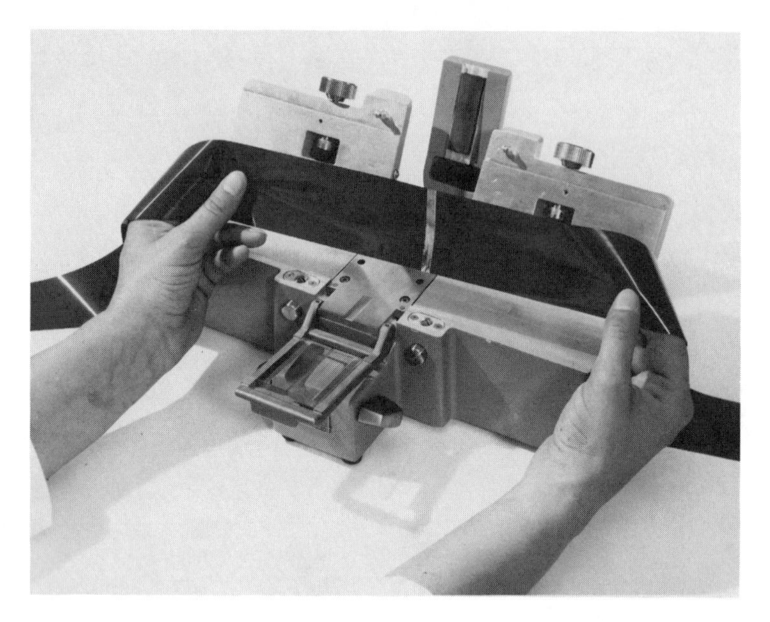

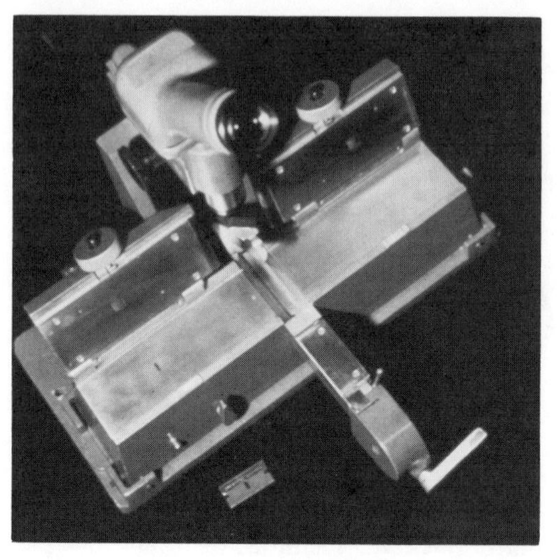

FIGURE 1.11 (above)
Completed videotape splice on two-inch tape. (Courtesy Ampex Corporation)

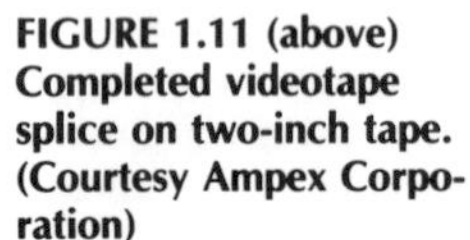

FIGURE 1.12 (above right)
Smith splicer with 40-power viewing scope.

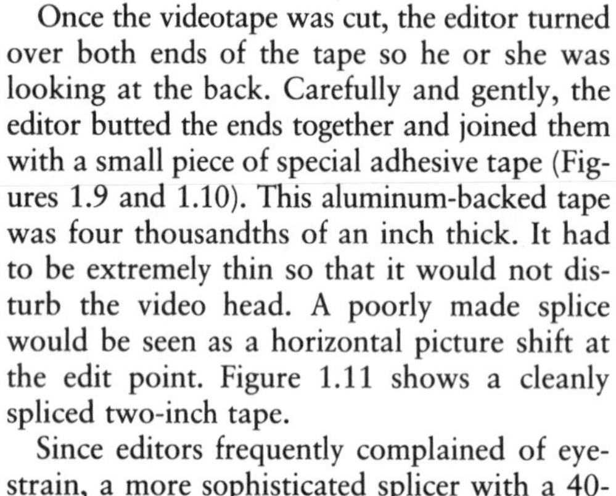

Once the videotape was cut, the editor turned over both ends of the tape so he or she was looking at the back. Carefully and gently, the editor butted the ends together and joined them with a small piece of special adhesive tape (Figures 1.9 and 1.10). This aluminum-backed tape was four thousandths of an inch thick. It had to be extremely thin so that it would not disturb the video head. A poorly made splice would be seen as a horizontal picture shift at the edit point. Figure 1.11 shows a cleanly spliced two-inch tape.

Since editors frequently complained of eyestrain, a more sophisticated splicer with a 40-power microscope attached was developed (Figure 1.12). Doors to hold the tape in position and precision rubber rollers also were added to move the tape in one thousandth of an inch increments. The trusty razor blade was replaced by a precision guillotine cutter. This new splicer made possible repeatable precision splicing.

From 1958 to 1968, I estimate that I made more than 20,000 physical tape edits and wore out enough razor blades to keep the manufacturers in business. By 1970 more than 300 television programs were edited at NBC, Burbank, in this manner.

FIGURE 1.13
The RCA TRT-1 color videotape recorder.

THE BEGINNING OF ELECTRONIC EDITING

In the next few years videotape recording techniques improved rapidly, resulting in noticeable improvement in both picture and color quality. Many more advertisers were asking to have their programs broadcast in color on a regular basis.

Before 1961 nearly all television receivers and broadcast equipment were built with vacuum tubes. These tubes were bulky, consumed a great deal of electricity, and generated a lot of heat. One RCA color videotape recorder, the TRT-1, contained almost 500 vacuum tubes (Figure 1.13). In 1961 RCA brought out the first fully transistorized videotape machine, the TR-22. It was much smaller and far more efficient than its predecessors. A short time later, Ampex introduced its own fully transistorized machine, the VR-2000.

In the early days of videotape editing, the

only method of making an edit on videotape was to physically cut and splice it. In 1963 Ampex brought out Editec (Figure 1.14), the first commercial electronic videotape editor. A small computer enabled the user to program edit points with frame-accurate precision.

By pressing a button on a control panel, the editor could record a single-frame audio tone on the secondary cue channel of two-inch videotape. This tone was used to establish a beginning or end edit point, and if the point selected was incorrect, the tone could be rerecorded on the desired frame.

Although this process was time-consuming, Editec helped open up a new world of editing techniques. It eliminated tape handling, physical splicing, adhesive tape, and razor blades. Electronic editing eliminated the need to cut the original, and the edited master was created by electronically splicing together scenes transferred from one VTR to another. If you made a mistake while making an electronic edit, you merely remade the edit.

This process is known as transfer editing and is how all electronic editing is done today. Even though electronic editing creates an edited master tape one generation away from the first, the original production material is untouched. Other versions can be created from the original. Figures 1.15 and 1.16 show two electronic editing systems used during the 1970s.

Another important feature of Editec was the ability to animate on a frame-for-frame basis. Although this option is not used much any-

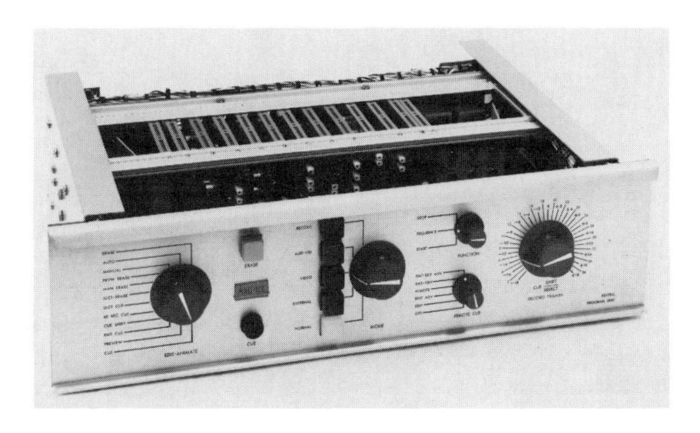

FIGURE 1.14
Ampex's electronic Editec editor. (Courtesy Ampex Corporation)

FIGURE 1.15
EECO stand-up two-inch videotape time code–based electronic editing system. (Courtesy NBC)

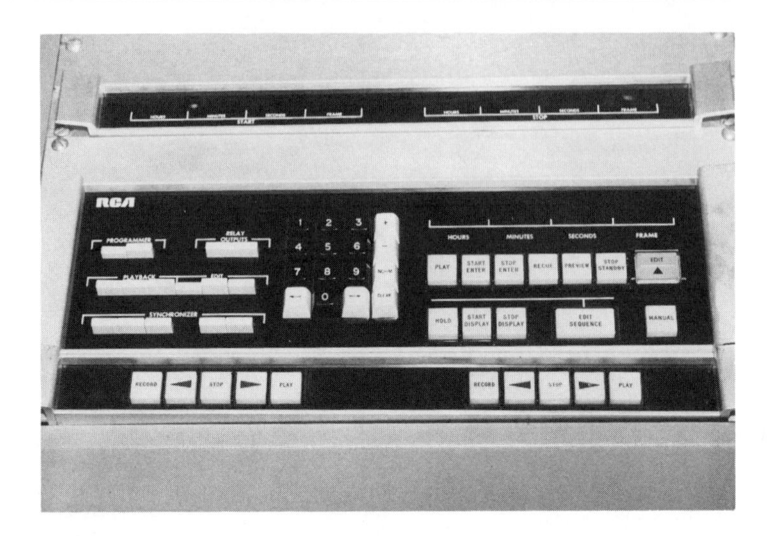

FIGURE 1.16 (above) RCA's Electronic Editor control panel. (Courtesy NBC)

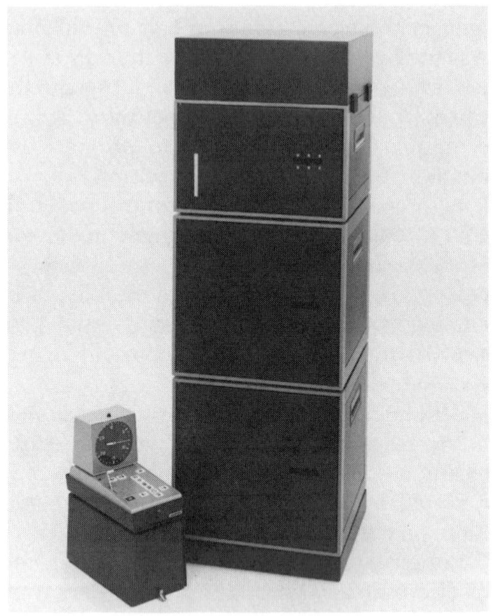

FIGURE 1.17 (above right) The Ampex HS-100 slow-motion videodisk recorder. (Courtesy Ampex Corporation)

more, at the time it provided a flexibility on tape that up to then could be achieved only on film.

As mentioned earlier, low-band color edited masters were of insufficient quality to be duplicated for syndication. To overcome this problem, Ampex introduced high-band color recording in 1964. Copies of more than ten generations could be made with far less degradation than before. Not even motion picture film could be copied with this kind of quality.

THE FIRST ELECTRONIC SPECIAL EFFECTS

Videotape had many other uses. One of these was the instant replay during sporting events.

FIGURE 1.18 The Ampex HS-200 production console was a companion unit to the HS-100 videodisk recorder. (Courtesy Ampex Corporation)

NBC was the first network to provide instant replays in color. I participated in the first attempt during the Rose Bowl on New Year's Day 1965. We were able to provide real-time instant replays in color but not freeze frames or slow motion. The videotape operator recorded each play in real time, then quickly rewound the tape and upon the director's command replayed the action seconds after the live action had been completed. In most cases, replays during sporting events occurred within ten seconds of the completion of the play.

In 1967 Ampex introduced the HS-100 videodisk recorder (Figure 1.17), which could record and play back up to 30 seconds of high-band color information, as well as provide freeze frames and slow motion. ABC used the HS-100 during its coverage of the 1968 Olympic Games, and the HS-100 soon became an important tool in electronic editing and post-production. The Ampex HS-200 production console (Figure 1.18) was used in conjunction with the HS-100.

That same year helical-scan VTRs appeared. (See Figure 1.19.) Their cost was relatively low, and consumers could now afford videotape equipment for home use. Without going into great technical detail, I should point out that a helical-scan VTR is a video and audio recording device designed to record and play back audio and video signals from half-inch magnetic videotape. The term *helical scan* refers to the method used to thread the tape around the video head in a helix, or spiral, pattern. (See Figure 1.20.)

More specifically, the video signal is recorded on the tape at a low angle slant as opposed to the nearly vertical video tracks recorded on two-inch videotape. For this reason, helical-scan recordings also are referred to as slant-track recordings. Nearly all VTRs built today for broadcast and consumer use are of the helical-scan variety.

In the early years of helical-scan recording, the tape was wound onto open reels and had to be threaded onto the VTR by hand. Today, except for professional one-inch broadcast VTRs, virtually all VTRs designed for the professional or consumer market use a tape that is contained in a plastic cassette. These are called videocassette recorders (VCRs).

Half-inch helical-scan recorders of the open-reel variety were used extensively in electronic editing systems for educational and commercial applications. The major problem with these recorders was that they were compatible only with other recorders of the same make. Only with the advent of three-quarter-inch VCRs in 1973 did companies agree on a standard format that would allow tapes made on one machine to be played on a machine from another manufacturer. The format recognized by all VCR manufacturers is the U-matic format pioneered by Sony. This was a giant step forward because it eventually led to the development of sophisticated editing systems using videocassettes.

Advances in electronic technology came fast and furious in the 1960s and 1970s. In 1967 the Electronics Engineering Company (EECO) of California developed time code for electronic editing (Figures 1.23 and 1.24). EECO added a form of clock time to the secondary, or cue, channel of the videotape. This time code allowed precise editing with greater flexibility than had been available in the past. EECO was awarded an engineering Emmy in 1971 for this

FIGURE 1.19
The Ampex VR 7800 one-inch helical-scan VTR using the now obsolete A format configuration. (Courtesy Ampex Corporation)

FIGURE 1.20 (below right)
The Ampex helical-scan VTR showing the helix threading path. (Courtesy Ampex Corporation)

FIGURE 1.21 (below left)
The Sony 2860 three-quarter-inch helical-scan VCR.

FIGURE 1.22
The Sony 2860 series helical-scan VCR. (Courtesy Sony Corporation)

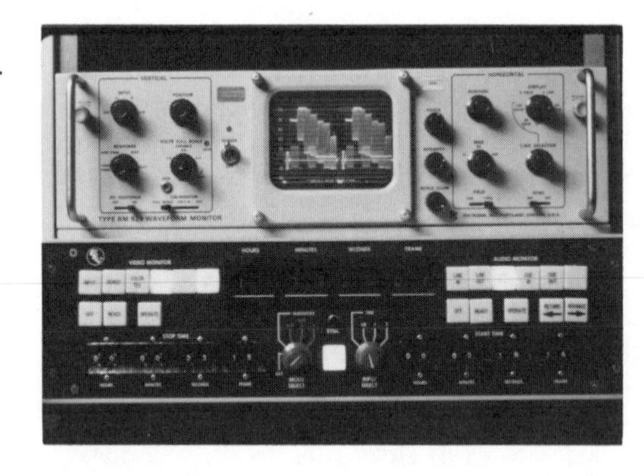

FIGURE 1.23
The EECO electronic editing control panel.

FIGURE 1.24
The EECO stand-up two-inch electronic editing system.

development. A few years later, this time code was improved, and a new compatible standard was adopted by the television industry.

Also in the early 1970s, Ampex introduced the AVR-1, a dual standard low-band and high-band VTR (Figure 1.25). It had many sophisticated features not found in VTRs of the day. For one thing, it was able to wind and rewind videotapes more than twice as fast as other VTRs. In addition, a stable picture appeared almost the instant you touched the play button. (The picture on older VTRs took three to four seconds to became stable.) The time base corrector (TBC) in the AVR-1 used the latest technology in digital electronics to eliminate electronic errors such as banding, which is seen as uneven levels of color from head to head forming bands of color across the picture. The TBC was not a cure-all, but the picture quality was noticeably improved.

Shortly after Ampex came out with the AVR-1, RCA introduced the TR-70, a videotape machine with excellent picture quality. Although compatible with the Ampex machine, the TR-70 was based on a different design. Both the TR-70 and the AVR-1 were used extensively in the broadcast industry until the early 1980s.

A two-inch version of the helical-scan VTR was introduced by the International Video Corporation (IVC) in 1973. The IVC 9000 used essentially the same recording method as other smaller format helical-scan recorders, but because of specialized electronic circuitry, the picture quality was superior and duplicates could be made with little degradation. Although not compatible with two-inch quadraplex tape recorders, the IVC 9000 was used extensively in the educational and industrial markets and to archive television programs.

In 1973 RCA won an Emmy for the development of the TCR-100, an automatic VTR that used cartridges to record and play back commercials and other short segments for airing. The TCR-100 alleviated many of the problems associated with the manual operation of the VTR and saved stations thousands of dollars in rebates to advertisers due to incorrect airing of commercials. Ampex later developed a similar machine, the ARC-25.

In 1974 the RCA TPR-10 made its debut. A portable two-inch broadcast recorder, this VTR found wide acceptance in the television market. It had many features not found on other recorders of this type. It was originally designed for the military to be used in jet fighter planes, but a modified version was made available to the television industry. Also in 1974 RCA introduced another VTR called the TR-600. This lightweight, low-cost tape recorder found many applications at small television stations, and a modified version was used in studio recording and post-production editing rooms.

Around 1977 one-inch broadcast VTRs began replacing two-inch VTRs in broadcast and editing applications. By 1987 virtually all video production was being done on one-inch or smaller format high-quality VTRs. Some two-inch VTRs are still used for duplication, since many small television stations use two-inch VTRs for broadcasting syndicated programs. By the early 1990s two-inch VTRs will no doubt be a thing of the past.

FIGURE 1.25
The AVR-1 two-inch VTR. (Courtesy Ampex Corporation)

TAPE VERSUS FILM IN TELEVISION

By the time commercial television arrived in the early 1940s, the motion picture industry had had a tremendous impact on the entertainment habits of millions of people. Television began to change these habits, but the motion picture industry continued to flourish not only in its own right but also as a production medium for television. Stars such as Lucille Ball, Joan Davis, and Dick Van Dyke filmed their television series, and many more followed. Since the mid-1960s, about 80 percent of prime-time programming has been produced on film.

Although film was prominent in the 1950s and 1960s, when videotape arrived in 1954, the trend in television production began to change. Fred Astaire was one of the first stars to videotape his television specials, and Bob Hope and Danny Thomas soon followed. The term *stop-and-go recording* was coined in the early days of tape editing to denote that a television program was not being shot straight through, but was being recorded in many small pieces that would be edited together.

Many different techniques were used in the early days of videotape editing, but one of the most successful was the method developed at NBC in Burbank, California, in 1957. This was really an adaptation of motion picture editing techniques. Basically, a film or kinescope work print was made by playing back the program material shot on videotape and photographing the picture from the face of a special kinescope picture tube (Figure 1.26). This created a 16mm black-and-white work copy along with a separate 16mm magnetic sound track containing the program sound, both recorded at the standard 24 frames per second. The kinescope work print was edited using conventional film editing equipment (Figure 1.27).

Engineers at NBC developed a special guide track known as the edit sync guide (ESG) or the "talking clock." The numbers on the kinescope work print and the videotape coincided frame for frame when the 30-frame kinescope was developed. By listening to these numbers on the optical sound track of the work print film and finding the same numbers on the uncut videotape, the editor could cut the videotape quickly and accurately.

FIGURE 1.26
A 16mm kinescope camera used to make film work prints for editing videotaped programs.

FIGURE 1.27
Conventional Moviolas used to edit kinescope work prints that will later be used to conform the original videotape.

A new era in television entertainment began on September 9, 1967, when NBC aired "Rowan and Martin's Laugh-In Special." The following February "Laugh-In" became a regular series. Each show averaged 350 to 400 tape splices. It took 50 to 60 hours of mechanically splicing the tape together to build each edited master. In addition, at least 30 hours of editing were required to build the work print, which the producer had to approve before the tape could be cut.

Figure 1.28 shows the preparation of an edit log by reading and logging the ESG guide track from the edited kinescope film work print. Figure 1.29 is the actual log used to edit the videotape on the "Laugh-In Special."

"Laugh-In" made editing history in many ways. It was the first extensively edited network show that appeared on a weekly basis. It also set a new editing trend that has been copied many times over the years.

"Laugh-In" was a proving ground for many new editing ideas, but one major disadvantage of splicing the tape together was that only cuts could be made. No dissolves, wipes, or other special effects could be used unless they were created ahead of time and inserted in the program as completed optical effects. To produce these effects on videotape, the editors of "Laugh-In" would have needed a computerized editing system such as the CMX-600, introduced by the CMX Corporation in 1971. A switch to such a system, however, would have been too impractical and costly for the show.

The CMX-600 (Figure 1.30) was capable of storing about 27 minutes of raw material in random form from which various sequences could be built. (See Figure 1.31.) Even though the picture was in black and white and the quality was mediocre, the editor could access any scene instantaneously, which allowed him or her to concentrate on the aesthetics of editing rather than the mechanics. Another amazing feature was that the pitch of a person's voice remained the same at any speed, even up to ten times normal playing speed. This feature was important to the editor because it eliminated two problems caused by conventional methods of editing either film or tape. In conventional methods, at speeds slower than normal play, the pitch of the sound becomes progressively lower as the speed of the film or tape sound is

FIGURE 1.28
Creating a log from the edited kinescope film work print that will be used to edit the two-inch master videotape.

AIR COPY VTR	9	NBC BURBANK VIDEO TAPE EDITING DEPT. CUE SHEET	24 FRAME KINE COUNT	X
PROT. COPY VTR	11		30 FRAME TAPE COUNT	
COLOR X	B/W		EDITOR ART S.	

REC. DATE 7-18-67	PLAY DATE 9-9-67	DATE 8-10-67	PAGE 2	OF 14
REEL # 1	PROGRAM ROWAN MARTIN'S LAF-IN	ESG COUNT BY ART S.	CHECKED BY W.K.	

	ESG CUE MIN	SEC	FRAME	CUTTING CORRECTIONS	TOTAL	SCENE DESCRIPTION	SCENE TAKE	LOAD NO.
1	02	39	10	X -1/4		MONOLOGUE		4
2	06	41	16	X -1/4			1	
3	17	31	00	X 0		COCKTAIL PARTY		2
4	17	50	00	X -1/4			T-3	
5	03	28	22	X -1/4				2
6	03	52	08	X -1/4		COCKTAIL PARTY INSERT	T-2	
7	32	07	13	X +1/4		white Eye		2
8	32	13	00	X -1/4			20-1	
9	05	59	00	X 0		COCKTAIL PARTY	P16	2
10	06	18	14	X +1/4		PU	T-2	
11	29	28	01	X +1/4		PARTY		2
12	29	31	02	X +1/4		#1 DANCING INSERT		
13	21	16	05	X +1/4		COCKTAIL PARTY		2
14	21	31	21	X -1/4			T-3	
15	22	33	00	X 0		#1 "Very Interesting"		1
16	22	36	12	X -1/4				
17	10	13	05	X -1/4		PU. BARBARA F.	P16	2
18	10	20	18	X +1/4		PU	T-2	
19	22	04	21	X +1/4		COCKTAIL PARTY		2
20	22	12	15	X 0			T-3	
21	10	29	21	X +1/4		MONTY	P16	2
22	10	56	23	X 0		PU	T-2	
23	22	42	10	X 1/4		COCKTAIL PARTY		2
24	22	57	07	X 0			T-3	

C5-65-as rev.

FIGURE 1.29
An editing log from "Laugh-In" used to cut the two-inch videotape.

decreased, making the task of finding sound edit points difficult and time consuming. As the tape or film sound speed is increased beyond normal play speed, the sound is turned into jibberish again, making the sound unintelligible. The CMX-600 system eliminated these problems by allowing the editor to listen to sound tracks at any speed from still to ten times normal play speeds while maintaining the correct pitch of the sound as though it were being played at normal speed.

Figure 1.32 is a typical CMX-600 equipment room showing the computer, six magnetic disk drives, and the associated support hardware. Each disk drive contains 4½ minutes of storage space, but the six drives work in conjunc-

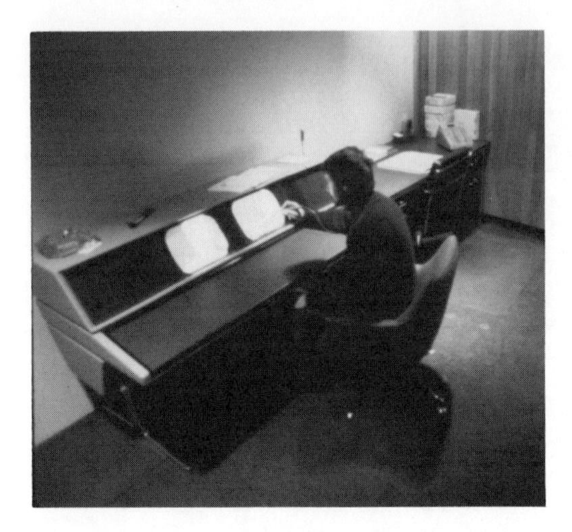

FIGURE 1.30
The CMX-600 disk-based editing system console.

FIGURE 1.31
Editing on the CMX-600.

FIGURE 1.32
A CMX-600 equipment room showing six disk drives on the left and disk storage space on the right.

tion with each other, acting as one mass storage device containing a total of 27 minutes of material.

Soon after the development of the CMX-600, the CMX-300 was born (Figure 1.33). This was known as an on-line editing system (see Chapter 6) and was used to build high-quality edited videotapes for broadcast applications. Newer versions of this editing system are in use worldwide today. Some people consider the CMX line of editors to be the industry standard.

CONCLUSION

As you can see from this introductory chapter, the production of film and videotape has come a long way in this century. In the following chapters, I describe the post-production process in greater detail. Film and tape editing has been my love for more than 35 years, and I hope to instill that same feeling in you as we explore this field together.

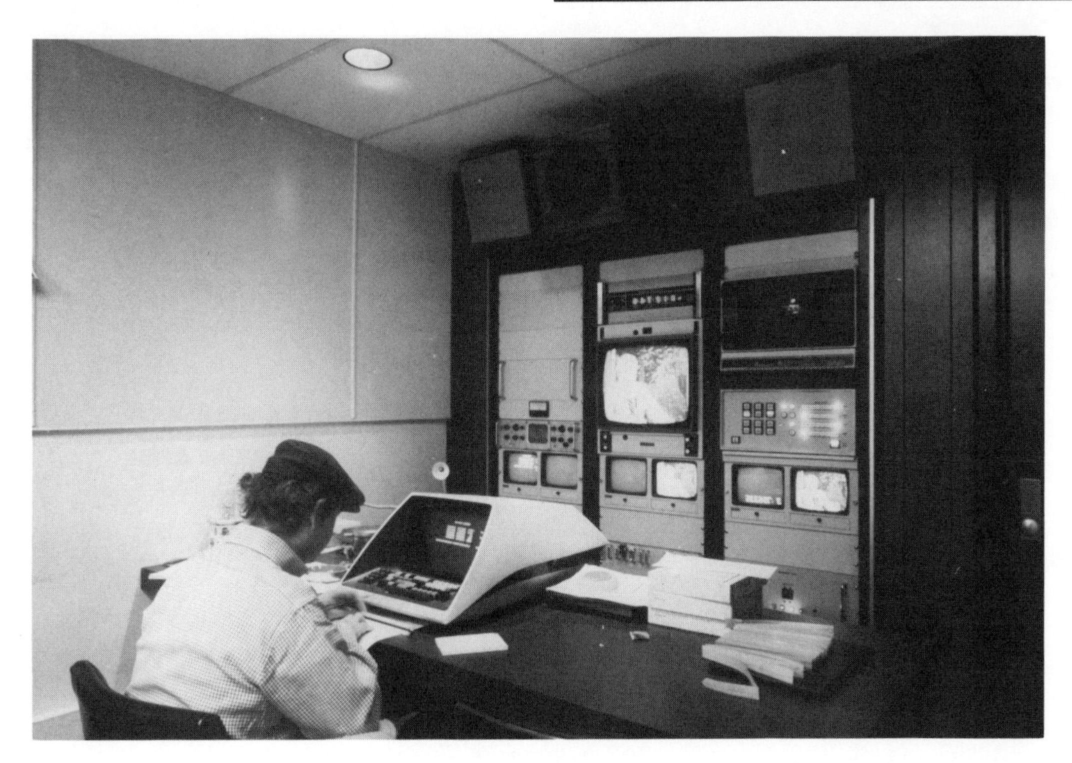

FIGURE 1.33
The CMX-300 on-line two-inch videotape editing system.

Why Tape? 2

While working as a film editor at NBC in Hollywood in the mid-1950s, I had the opportunity to get involved in what I felt was a new and exciting challenge. Videotape had arrived. Even though earlier experiments by Bing Crosby Enterprises had proved somewhat successful, the design of videotape equipment was not considered practical for the television industry. That all changed in 1956 when Ampex unveiled the first broadcast black-and-white two-inch VTR.

At first, many people thought that videotape was just a toy since it could not be edited and you could not see images on the tape when you held it up to the light. But the fact that you could place a roll of two-inch blank tape on a machine, push a button, and record pictures and sound intrigued many of us. To be able to rewind the tape and see and hear what you recorded moments before was even more intriguing.

Those early VTRs were crude by today's standards, but within a year, we were successfully editing videotape and putting together relatively complex television shows quickly and inexpensively.

In the early days of videotape, few people were qualified to deal creatively with this new medium. Today, more than three decades after the first videotape machine was unveiled, the videotape industry employs thousands of people who are trained to use the most sophisticated equipment available.

Videotape editing today is a far cry from the days of the razor blade and microscope. Today's minicomputers are so powerful and small that the computer and its memory are built on a single board weighing less than one pound. On that board are stored thousands of instructions, including many hundreds of editing decisions. Computer assisted editing systems perform a multitude of tasks too tedious for most editors to do by hand. This allows the editor to think creatively rather than mechanically.

Videotape images are currently recorded using a method known as analog recording. This method results in image degradation with each successive generation, or copy. The wave of the future is digital recording, a method that does not degrade the image no matter how many times it is copied.

We can now record television signals in digital form, allowing us to duplicate hundreds of generations of images and sound with no noticeable loss of picture or sound quality. Although complete programs are not yet recorded digitally, by 1990 this technique will begin to replace the current analog method of recording television signals. Digital video and audio will give the home viewer increased picture and sound quality and give the editor increased flexibility.

Digital technology lets us take apart the video image. We can squeeze it, make it smaller or larger, zoom it, twist it, split it apart, spin it, flip it, turn it upside down and backward, or manipulate it in just about any way. And we can do this in a matter of seconds or minutes.

Highly complex optical effects are easily accomplished on digital systems. How much nicer it is to sit in a comfortable tape editing room and see the results of special effects instantaneously than to send them out to be created at a film lab while you wait days or weeks for the finished product, uncertain whether the effect will turn out the way you visualized it.

WHY FILM?

Film can't touch tape when it comes to special effects, so why are more prime-time shows produced on film than on tape? There are a number of reasons, but one of the more important is that people used to working with film like the look of film. To film people, the videotape image looks flat.

The primary difference between a television camera picture and a film image seen on television is that the television signal is made up of pure colors based on the additive color system using the three primary colors—red, green, and blue. The film image is created using dye images based on the subtractive complementary colors yellow, magenta, and cyan. The film image has more contrast than the tape image, although both images might look good to the viewer by themselves.

Another reason why so many shows are still produced on film is that many film people are afraid of tape. Many people trained in film either have not taken the time to study tape or have had a bad experience with tape. Some of the reasons film people use for avoiding tape are "Jobs will be lost to videotape," "I can't deal with something I can't see," or "I'm intimidated by computers."

When film people give videotape a chance, they often find that not only are the terms used in the two media very similar, but also many technical positions in film parallel those in tape. Many positions in tape production require little or no additional training compared to their film counterparts. Others require training only insofar as they must learn to operate some tape-specific equipment.

UTILIZING FILM IN TAPE PROGRAMS

The Process

Although editing on film can be an extremely creative process, it also can be slow and cumbersome. (See Figure 2.1.) The moviola (a device used to edit film) cannot be used to judge the efficacy of dissolves or other special effects. Film technology requires that any effects be made in advance and cut into the film work print at a later date. This inhibits aesthetic decisions and slows down the creative process.

FIGURE 2.1
A 35mm flatbed film editing machine.

In addition, scenes that are shot today cannot be viewed until tomorrow because the film must be developed and printed and any sound must be electronically printed and synchronized with the picture. If there is a question about lighting, makeup, camera angle, or some other production problem, the director can either keep doing takes until he feels he has captured what he wants on film, or he can wait until the next day to view the film and hope he got it. If not, he will have to take time out of that day's production schedule to do retakes, which is not only time-consuming but also expensive.

Even though working with videotape might eliminate many of these problems, some people still prefer to work with film. Instead of using film for the entire post-production process, however, they incorporate the two media to get the best of both worlds.

Many television film producers shoot on 35mm color negative film, transfer the content to videotape, and edit and release the finished product on tape. Chapter 10 outlines this process in detail.

For productions that require a film release, one of two approaches may be taken. In both cases, the filmed content is transferred to tape to be edited. Once it is edited, the producer can either conform the original camera negative according to time code data and film edge numbers derived from the edited tape or convert the edited videotape directly to 35mm or 16mm color film. The second approach results in a lower quality image because the image on videotape that is being transferred to color film through an electronic system is at least of second or third generation.

The first method, which is too complex to discuss in detail here, actually conforms the camera negative, which is printed by conventional means and released as a high-quality film product. If any complicated optical effect was generated in the videotape version, it will be very costly to duplicate it in the film domain. The second method has the advantage of including all the special optical and digital effects created during the videotape editing session, but the converted film image is noticeably degraded by the electronic process.

Videotape technology also might be useful in making feature films for theatrical release. Using a computer assisted videotape editing system can cut post-production time by as much as 50 percent. The process of shooting on film, editing on tape, and releasing on film could be used to good advantage in making feature films as well as television programs.

The People

In 1975 the Motion Picture Film Editor's Guild in Los Angeles bought a computer assisted editing system for the sole purpose of training their members (who are already trained film editors) in videotape editing. Working as one of the instructors, I found that the majority of my students viewed this equipment as just another tool. Some people, however, rebelled at the thought of editing by computer. Fortunately, those students were in the minority.

The Editor's Guild training course is designed to acquaint film editors with the computer keyboard and to familiarize them with how to make edits and create optical effects. It is not designed to turn them into full-fledged tape editors. There are, however, advanced courses designed to let editors practice what they have learned until they become proficient.

Film-style videotape editing machines such as the Ediflex, Editdroid, and Montage are making it easy for film editors to make the transition to videotape. In addition, videotape editors are now being treated with more respect. In the past, many people in the film industry considered those involved in videotape as engineers and technicians with only a thimbleful of creativity. That attitude is gradually changing.

It is wrong to assume that film and tape techniques are identical in all areas. As pointed out earlier, many functions in film can be directly related to those in videotape. But it is important that film people consult with qualified and experienced videotape people before going into their first videotape production. Doing so can prevent costly errors.

One important point to remember is that videotape will not replace motion picture film, at least not in the near future. For one thing, the resolution of motion picture film for theatrical use far exceeds that of videotape. And unless some miraculous breakthrough occurs in television production and broadcasting, the present U.S. broadcasting standards will be with us for a long time. As this chapter points out, however, many filmed television programs and features could be done on videotape more quickly and more economically. Those two considerations, along with today's technology and the growing number of creative and talented videotape editors and technicians, make videotape a viable alternative for all types of productions.

The Videotape Post-Production Process

3

In this chapter, I outline the sometimes mysterious and often challenging process called post-production. I present the process chronologically from the time the material has been recorded on tape and is available to the post-production people, through the various editing steps, to the finished product that is ready for distribution or broadcast.

BASIC EDITING HARDWARE

The basic editing system consists of two VCRs—one player and one recorder/editor—and some sort of device to control them, along with picture and sound monitors. All the equipment must work together well, or you will find the editing process unnecessarily difficult and frustrating. Figure 3.1 shows a typical off-line editing setup.

Chapter 6 discusses editing systems in detail, so I will limit my discussion here to the main features required in a system. A VCR designed for the consumer market generally records and plays back tapes but cannot be used for editing, as that requires remote control capability. Only industrial models or the very high end consumer VCRs are equipped with the remote controls that are required for use with electronic editing systems. Furthermore, for a VCR to edit material together without breakup or other disturbances at the edit point, it must be capable of electronic editing. Thus any half-inch VHS or Beta machine, 8mm video recorder, or three-quarter-inch U-matic machine can be used with any edit controller system as long as it meets these requirements.

The second requirement is that the manufacturer of the edit controller must provide a piece of hardware that allows the controller to communicate with the VCR. Without this interface, the controller will not be able to control the VCR, even if the latter has remote capability. This is generally not a problem when you are renting a system, but it is an important consideration when you are buying an editing system.

Some editing systems have as many as six VCRs and generally use the three-quarter-inch format. When more than one source (or play) VCR is used, the amount of hardware required and the cost rise dramatically.

When editing on systems with two or more source VCRs, you will need a time base corrector (TBC) to produce optical effects such as fades, dissolves, and wipes. This device is used to stabilize the picture and synchronize the source machines so that a smooth, jitter-free picture results during an optical transition. Other equipment that works with the TBC as well as time code reading hardware also are required.

There are two methods of editing in use today. The first is called control track editing and is used on inexpensive editing systems. In this method, edit accuracy is based on counting control track pulses, which are analogous to film sprocket holes. This method of editing is neither repeatable nor always frame accurate.

The other method of editing uses time code. If the hardware and software of the editing system are operating correctly, this method is

FIGURE 3.1
A typical off-line editing setup.

frame accurate on a repeatable basis. Time code editing is generally found on the more expensive editing systems. Briefly, time code is a method of indexing every frame of videotape with a unique address, or location.

The term *invisible time code* refers to the longitudinal time code recorded on a track separate from that on which the program sound was recorded. Invisible time code can be displayed only by using a special time code reader. This type of code is used by computer assisted editing systems to make frame-accurate edits. Time code is explained in detail in Chapter 5.

BASIC EDITING PROCESS

The basic videotape editing process is as follows:

If possible, have the editor in the production meeting to help avoid problems in post-production.
Record the live material on tape along with any film transfers.
Make a work copy of the tape.
Review the material and select good takes.
Generate a paper edit list.
Perform the off-line edit.
Get the producer's and/or the director's approval.
Clean the edit list.
Make any editing changes.
Perform the on-line edit; build the master edited tape.
Add digital effects.
Add graphics and titles.
Color-correct the picture.
Add sound enhancement.

These fourteen steps represent every phase of the editing process through completion of the finished master ready for release. Once the edited master has been completed, the finished product can be duplicated and syndicated.

Although I discuss all fourteen items, not all of them will be used in every situation. For example, rather than creating a work tape first, some producers build a master edited tape using original production material.

Several production people might be involved in the editing process, including the director, the producer, the executive producer, and the network representative if the program material is to be broadcast. The director usually works directly with the editor, but in some cases he or she might have the assistant director or a production assistant assemble the material with the editor.

Generally, the director makes the first cut. The producer might then add his or her input. The executive producer has the final say over the end product. If the program material is going to be broadcast, a network or station representative from the standards and practices department might inject the broadcaster's viewpoint.

OFF-LINE OR ON-LINE EDITING?

Off-line literally means "away from the terminal," and on-line means "at the terminal." How do you tell whether the editing process will be off-line or on-line? Whether the tape format is quarter-inch, half-inch, three-quarter-inch, or one-inch, the final use of the finished product determines what editing process will be used. Generally, however, any videotape that is acceptable for viewing or broadcasting and is of sufficient technical quality to be duplicated could be edited by the process known as on-line editing, whatever its end use might be. Alternatively, a videotape with time code numbers imprinted on the picture area (called a window dub), such as a tape produced in an off-line editing session, is not considered suitable for anything other than checking the continuity of the material.

On rare occasions, a window dub (Figure 3.2) might be the only original source material available. One example of this is a documentary in which certain footage is available only in work print form. Another example is when outtakes of a program are used for their entertainment value in other programs. In this case, the producer might have no other choice but to use the imprinted tape. Sometimes an editor will try to mask the time code with a block of color similar in density and shade to the objects around it. It is not, however, a good idea to use this type of material unless it is absolutely necessary.

THE PRELIMINARIES

Recording the Material

The program material can be generated in one of three ways. The first is by photographing scenes with a television camera in real time. This is called live recording. A second method is to transfer material on motion picture film to videotape by means of a special piece of equipment known as a telecine film device. Briefly, a telecine machine is a color camera and special projector connected together so that the camera looks at the film on a frame-by-frame basis, transforming it into an electronic signal that is then recorded on videotape. Finally, material might be introduced from electronic equipment designed to generate artwork, graphics, titles, or any other type of material not recorded live with a video camera.

Arranging a Post-Production Meeting

Arranging a post-production meeting prior to the tape session is advisable because the editor might be able to offer suggestions that will make the editing process run more smoothly. Generally, post-production meetings help the director avoid situations that might cause problems in the editing room. If there is no post-production meeting prior to taping, the editor can only hope that the director knows what he or she wants and has shot the material in such a manner as to allow it to go together efficiently. If this is not the case, the editor might have trouble making good edits or generating the desired special effects.

THE OFF-LINE EDIT

There are three possible reasons for skipping the off-line editing process. They are time, money, and the need for complex special effects. In some cases, the time available for editing does not allow the producer the luxury of making work copies from which the usable material can be culled. In some situations, the budget might not allow for the cost of making work tapes and paying for the additional off-line editing time. Finally, it is not cost-effective to build special effects in an off-line editing session because most off-line editing systems are not equipped to do this. Building these effects in an off-line session requires the use of expensive switchers and digital effects equipment, and the effects would then have to be duplicated in an on-line editing session. Alternatively, if the director chooses to skip the off-line process, all editing decisions must be made at very expensive on-line rates.

FIGURE 3.2
A window dub showing time code information added to the picture to assist in locating edit points.

Making a Work Copy

Assuming that everything you need for an off-line session has been recorded on videotape, the next step is making a work copy of the original material. This is done for two reasons. The first is protection. Even though today's videotape recorders and players are designed to handle tape gently and reliably, there is the chance that a damaged reel or warped cassette will scratch or break the original tape and ruin sometimes irretrievable material. If a work tape is damaged, it is easy to make another copy from the original production material.

The second reason for making and using a work tape is cost. Although certain circumstances may dictate the skipping of the offline step, it is strongly recommended that it be used whenever possible for the following reasons. When you use the original footage in an editing session, you are actually doing an on-line edit, and the hourly rates for this type of editing might be three or more times those for an off-line session.

The work tape performs several functions. It can be used as a viewing copy to help the editor find the best take of a scene. By imprinting each frame with visible time code numbers, the editor can then generate a reference log. In addition, while the tape operator is making a work tape from the studio master, he or she can watch for defects or other problems that might influence the director to use a different take or even reshoot a scene if time and budget permit.

Sometimes a separate tape with window dubs is shot during the production. Although this can be helpful in compiling a reference log, any defects on the master production tape will not necessarily be detected because the window dubs appear on a separate tape and may not contain the same length of material due to being started later or stopped earlier than the master production material.

Having a work tape with window dubs facilitates the compilation of a reference, or edit, list. If the videotape player is equipped with a joystick, knob, or slider that controls the speed and direction of the tape, the editor can slowly move the work tape frame by frame in either direction, allowing him or her to jot down the time code location of each frame. (See Figure 3.3.) More complex editing systems allow the editor to type in the time code number of a particular frame and put the computer in search mode. The computer will automatically find the desired frame and make the edit.

Reviewing the Material

Work tapes can be viewed in the office, at home, or in some other environment that allows group decision making. Using work tapes to select the material for the finished product is ideal because you can view the same scenes over and over again. In addition, by viewing them in a private setting, you do not feel the pressure you might if you were paying an hourly rate for an editing room full of expensive equipment. Many directors prefer to take a cassette home for review and then come into the office with preliminary editing notes.

Generating a Paper Edit List

A paper edit list is an inexpensive way for you to sift through large amounts of material and select the scenes you might want to use, which are then recorded on paper. When a single video frame is frozen on the screen, you can easily log the frame's eight-digit time code number.

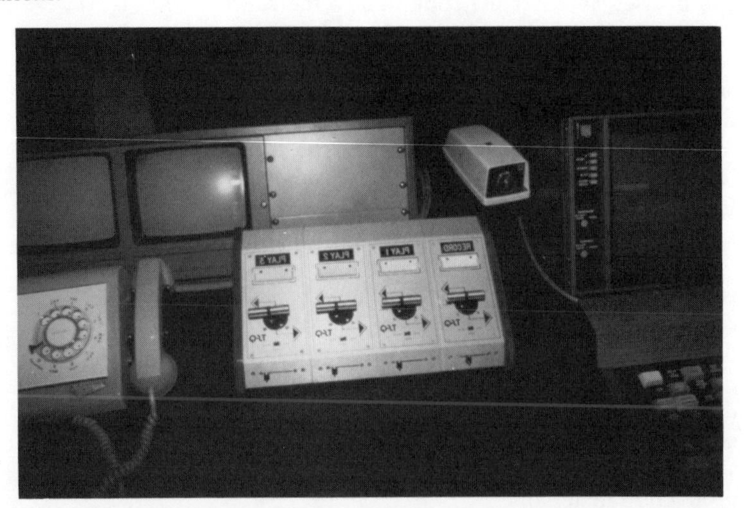

FIGURE 3.3
This joystick VCR motion controller allows the user to move tape remotely within a videocassette.

Two sets of numbers define the beginning and end of a segment of tape, which is generally referred to as a cut. Here is an example of a simple cut:

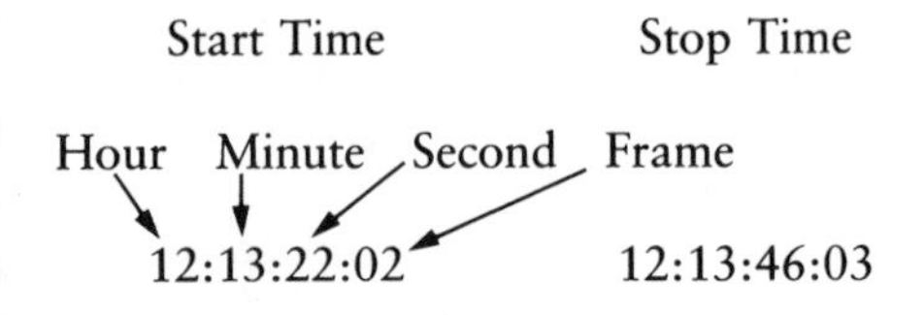

The length of this cut is exactly 24:01 (24 seconds and 1 frame), which is defined as the duration of the cut. An edit is made up of two or more cuts. Here is an example of a simple edit (two cuts):

Cut 1

Start Time	Stop Time
12:13:22:02	12:13:46:03

Cut 2

Start Time	Stop Time
07:11:09:04	07:11:19:10

Cut 1 has a duration of 24:01, and cut 2 has a duration of 10:06. A paper edit list is merely a way of recording a series of cuts for purposes of timing and continuity (Figure 3.4). This list need not be frame accurate because it is just a preliminary sketch of continuity and the approximate length of the program on paper.

Performing the Off-Line Edit

The next step is using the numbers on the paper edit list to create a rough assembly, sometimes called a rough cut. Even though the rough cut might not be frame accurate, it gives the director a visual idea of the continuity of the project.

A director who is very familiar with the material might skip the paper edit list and rough cut and go directly to the off-line editing process using more sophisticated editing equipment (Figure 3.5). This is practical only if the director has done his or her homework and is able to build the edited work print without having to make a lot of decisions about which takes to use.

Edit log

PAGE 1 OF 6
DATE FEB. 28, 1988

CLIENT: INDUSTRIAL CONSULTANTS
ADDRESS: 100 STATE STREET
TITLE: NEW PRODUCT LINE
PRODUCER: JIM SMITH
DIRECTOR: MARY BEAL
PHONE:

REEL	SLATE	HOURS	MIN	SEC	FRAMES	HOURS	MIN	SEC	FRAMES	COMMENTS
1	17-3	17	02	15		17	02	21		30 FRAME FADE-IN
6	114-1	03	14	14		03	15	00		FRANK'S INTRO A/V
			30 FRAME DISSOLVE TO							
3	26-1	09	16	04		09	16	22		VIDEO INSERT TO COVER
										BEGINNING OF NEXT CUT
3	27-1	09	19	03		09	19	30	00	WATCH AUDIO
3	28-2	09	21	00		09	22	07		HELICOPTER TAKE OFF
2	—	08	15	27		08	15	37		HIGH ANGLE OF CITY
2	—	08	15	49		08	16	03		ZOOM IN TO H.Q. BLDG.
2	16-1	08	03	03		08	03	10	00	C.U. DOOR
		30 DISS TO BOARD ROOM								
2	19-3	08	19	14		08	19	25	00	LS BOARD IN SESSION
4	20-1	10	00	14		10	00	23	21	CU DIAGRAM
			30 FRAME DISS TO							
4	21-1	10	01	17		10	01	29		INSIDE PLANT. NOTE
										DUB OFF/MAKE COPY
										FOR DISSOLVE.
4	22-2	10	02	03		10	02	33	16	SAM DISCUSSING
										PRODUCT
5	6-1	11	11	01		11	11	06		C.U. INSERT-BACK-IN
										FROM END OF CUT.
										SUPER OPENING
										TITLE. 30 FR. FADE
										IN, CUT OUT.

FIGURE 3.4
A paper edit log outlines general continuity and does not always have to be frame accurate.

Taking this a step further, an editor can also use the work tape to create optical effects. Based on time, the director's experience, and the budget, the editor might choose to use a sophisticated multi-VTR off-line editing system capable of producing optical effects such as dissolves, wipes, and fades. This allows the off-line editor to reach the polished edit stage much more quickly.

In this type of session, every edit is frame accurate. Dissolves and other effects appear as they will in the final product, and each editing decision is stored on either a floppy disk or paper punch tape, which allows each edit to be conformed later and duplicated on an on-line editing system.

Getting Approval

At some point, the producer, a network representative, or someone else with final approval over the continuity must see the work

FIGURE 3.5
The author is shown using a computerized off-line editing system.

print. This might occur after the director has finished with it or sometime during the editing process. The sooner the work print is approved, the easier it is to make changes and the less reediting is required.

Cleaning the Edit List

So far, the editor has built a work tape that represents the continuity of the project. During this process, many edits might have been made and stored in the computer's memory. These edits may include various attempts by the editor to make an edit work better. The next questions to be answered are "What do we do with all these numbers?" and "How do they relate to the final picture and sound?"

Since the late 1970s, computer editing software programs developed within the television industry have performed the highly complex computations required to clean an edit list before automatic assembly of the final version can be performed. Automatic assembly is possible without cleaning the list, but it is an expensive waste of time and money.

Two list-cleaning programs are very popular with editors using time code–based editing equipment. The first, called 409, was named after the household cleaner. This program removes duplicate edits, repositions out-of-sequence edits, fixes overrecordings, and solves a host of other technical problems that often occur in a computer-generated time code edit list. The second program, called Trace, traces time code edits through up to ten revisions and generates a single correct time code edit list that reflects the final edited work tape, which may be used to build a master tape by means of an automatic assembly. Thus, 409 is used to clean the list, and Trace is used to generate the final version of the list. A detailed description of 409 and Trace appears in Chapter 8.

THE ON-LINE EDIT

When all parties involved agree that the program has been off-line edited correctly, the edit list is turned over to the on-line editor. Sometimes the off-line editor and the on-line editor is the same person. This is generally the best way to work. Some editors work in teams, with one doing the off-line edit and the other the on-line edit. If both editors are aware of each other's problems, this can be a satisfactory arrangement.

While off-line editors generally learn how to use relatively simple computer assisted editing systems to generate a work tape and an edit list, on-line editors need a different kind of skill. They must know how to use complicated video switchers, digital effects devices, graphics generators, and a host of other technical equipment generally not available to the off-line editor (Figure 3.6).

In addition, the on-line editor must understand the video signal, how to read video display devices such as a vectorscope and waveform monitor, and how to use other complex devices that help ensure that the master tape will meet industry technical standards. The on-line editor must be able to operate sophisticated VTRs and know how to set them up properly for playback or recording. The editor must have enough technical knowledge to recognize a technical problem with the tape or the video signal.

The on-line editor must be able to hear distortion and judge the quality of the sound track associated with a video product. The editor also must have visual acuity so that he or she can adjust the color quality of the picture through sophisticated video devices such as time base correctors, image enhancers, and color correctors.

These skills are generally the result of years of experience as a videotape operator, but they also require a special creative ability. A good on-line editor must be both an artist and a diplomat who can produce a tape that meets the client's needs and expectations.

Adding Digital Effects

One of the more complex tasks an editor must master is the operation of a digital video effects (DVE) generator. Some digital effects devices can change a television picture in many different ways, twisting it, tearing it apart, zooming in on it, or modifying it in some other seemingly magical way. Specific effects include turning a flat image into a round globe or exploding an image into thousands of pieces and having them fly off the screen.

Using this type of equipment requires a thorough understanding of video switchers, storage devices, DVE generators, and other complex devices designed to manipulate the video image. The purchase price of a single channel of digital effects might cost anywhere from $40,000 to more than $300,000. Some complex effects require as many as four digital effects channels connected together to produce a single special effect.

FIGURE 3.6
A typical one-inch on-line editing room.

These effects are used to make a transition from one scene to another, enhance the video image, or produce some unusual effect not normally found in television production. Since these effects are rarely created in an off-line editing session, the off-line editor must indicate the beginning and end of the effect, as well as its duration, in the work print. This is usually done by substituting a simple dissolve or wipe to indicate the duration of the effect so the director will have some idea of the continuity of the edit. The off-line editor will usually include a note in the computer edit list detailing any special requirements needed to create such an effect during the on-line session.

Adding Graphics and Titles

A program is generally not complete without graphics and titles. These enhancements might be added when the off-line work print is being built, but often the producer does not have the necessary list of credits or know what graphics he or she wants to use until the on-line editing session. In addition, most of the sophisticated equipment used to generate these graphics is too expensive to be used in an off-line editing room.

Graphics and title generators might be one and the same. A few widely known generators are the Dubner, Aurora, Bosch FGS-4000 (Figure 3.7), and Chyron. These units are capable of generating graphics, letters, and drop shadows; colorizing images; and creating animation and three-dimensional objects. Some can digitize photographs and change the image into a different form.

One type of effect is posterization, which turns a live image into one that looks as though it was drawn by an artist. The Paintbox and Aurora are capable of airbrushing part of a scene so that you can add or delete part of an image. Suppose you wanted to remove a mustache from a man. A paint system would allow

FIGURE 3.7
The Bosch FGS-4000 Digital Graphics system. (Courtesy Robert Bosch)

you to erase the mustache and blend in the color and texture of the surrounding skin.

Specialized equipment to produce these effects is expensive and complex, but it offers editors an opportunity to do things that were impossible only a few years ago.

Color-Correcting the Picture

Once the on-line assembly begins, whatever is recorded on the master edited tape is what will appear in the final product. Therefore, the on-line editor must be sure that the video and audio signals are recorded with care and meet all the broadcast specifications of the network or television station airing the program. Among other things, the on-line editor must be able to adjust the color saturation (chroma), flesh tones (hue or color phase), video level (gain), and contrast (black levels) of the video signal.

Almost invariably, color bars are used as the reference for setting up videotape machines, color monitors, and any other equipment used to modify the video signal before it is recorded on the master tape. Without this or some other reference signal, editors have no idea whether the signal they plan to record on tape is the same signal they are viewing on the monitor and whether the video and audio levels are set properly.

At times during the off-line or on-line editing process, the signals coming off the source VTRs do not look quite right. This variation might be caused by lighting problems, improper video levels, or any of a number of other factors requiring adjustment of the color and/or video levels during the editing session. The simplest form of color correction for this type of problem is the color phase adjustment, or hue control, which rotates all the colors equally in the same direction. This control, along with several other controls, is located on the TBC, which is connected to the output of a videotape machine.

The TBC gets rid of electronic and mechanical errors in the videotape and videotape playback equipment that cause color errors, picture jitter, and other forms of visible instability. Although the TBC eliminates most of the technical problems, sometimes the color portion of the signal is so distorted that more drastic steps must be taken.

As mentioned above, the hue control shifts all the colors in the same direction. For example, if someone thought the flesh tones were a little too green, the tape operator would rotate the hue control away from the green and toward the red, which would make the flesh tones look more natural.

The monitoring device the operator uses as a reference to know how much to rotate the color and in what direction is known as a vectorscope. This device displays a pattern of the colors on a small, round picture tube with a green pattern on its face. The operator is able to see the colors rotate in a clockwise or counterclockwise direction and watch the visual effect on the picture by observing a calibrated color monitor.

A more complex and effective method of changing the hue is to insert a special color correcting device in the output channel of the playback VTR after the TBC. This device separates the color signals into their red, green, and blue components. The tape operator can then selectively reduce or increase any of the three primary colors individually to correct any deficiencies in the color portion of the picture.

When using this device, it is important to adhere to basic color standards and calibrations. If, for example, the color monitor that is being used as the absolute signal reference is out of balance, the operator will have no way of knowing whether the colors being recorded on the tape are the same as those being observed on the monitor. Therefore, the operator must calibrate all VTRs, color-correcting equipment, and monitors before attempting to adjust color and video levels.

Adding Sound Enhancement

Sometimes an off-line or on-line editor does not treat the sound track with the respect it deserves. Although there is little if anything the editor can do to maintain the quality of the sound track in the off-line editing session, the editor should listen for obvious distortion and other problems on the work tape. The off-line session is the time to track down the source of these problems. The editor should play the original production tape in a sound-controlled environment. If the problem is not on the tape, it could be the result of a problem with the VCR used in editing or a defective transfer. If time permits, a new work tape should be made from the original production material.

Stereo phase should be maintained in the off-line session, but it is a primary consideration during the on-line session. Out-of-phase stereo signals might even cancel each other out and result in no sound at all.

To understand phase, you must understand that there are two types of sound configurations. The first is called double-system sound, which means that the sound track is contained on a separate optical or magnetic material. This allows the editor to manipulate the sound easily with respect to its relationship to the picture. In the second configuration, called single-system sound, the sound track is part of the picture element. It is recorded on the same piece of videotape as the picture and therefore is somewhat more difficult to manipulate with respect to the picture.

True two-track stereo, whether single or double system, must be handled carefully to avoid sound cancellation and other problems associated with stereo editing. Monaural, or single-track editing, does not pose these problems.

The most important aspect of monaural sound editing is to maintain balanced levels, especially when dealing with two separate monaural sound tracks. For instance, one track might contain the production sound, while the other might contain narration, music, or sound effects. The proper balance must be maintained so that one track does not drown out the other.

In addition to making sure there are no problems with the sound, the editor might want to enhance it. In television this process is called *sweetening,* which is similar to the dubbing process in the motion picture industry. In sweetening, the editor blends music, sound effects, or narration with the production or dialogue track.

In most cases, once the edited master has been assembled, it is turned over to a facility specializing in sound mixing and enhancement. The person in charge of combining the tracks is the sound mixer.

The first step in the sweetening process is to put the production sound track on a multitrack audiotape recorder (ATR). ATRs usually have 2, 4, 8, 16, or 24 separate tracks on which the editor can record sound. To have sufficient flexibility, the sound mixer should use at least an 8-track ATR, but 16- and 24-track machines are more commonly used with two-inch audiotape.

One track of the ATR is reserved for time code, which most systems use to synchronize the sound with the picture. Another track is reserved for the production sound track as it has been edited in the on-line session. This track is transferred from the edited videotape master to another track on the ATR and becomes the basis for the final track after sweetening. This is called the laydown process.

After the main sound track has been transferred to the multitrack ATR, the next step is to add sound effects such as door slams and telephone rings not included during production or editing. The multitrack ATR is synchronized with a one-inch or smaller format VTR, usually a half-inch or three-quarter-inch machine. This tape is used for visual sync reference and has an eight-digit time code number imprinted in every frame. The VTR and the ATR are locked together electronically and synchronized by comparing the time code.

The mixer adds various sound effects stored on audiocassettes, cartridges, or reel-to-reel tapes. These effects are recorded on one or more sound channels and logged by the sound effects editor so the mixer will know where the sound is recorded and where it should appear on the main sound track.

Adding music to a program can be done in a number of ways. One is to purchase the rights to canned music from public domain libraries or licensed music companies. They charge you according to the amount of music you use or by the selection. You pay a royalty to them for the right to use the music. Another way is to have a composer write an original score or adapt already written music for your needs.

If an original score is used, the music is performed by a live orchestra and is recorded in one of two ways. First, it can be recorded "wild"—that is, without the conductor's viewing the picture beforehand. The conductor records the music to time, often by using a click track, which is a type of electronic metronome. Alternatively, the conductor might view the picture as he or she conducts the orchestra to make sure the music fits the picture. Whichever way the music is recorded, the music editor will then add it to one of the unused audio channels on the ATR. The process of laying down the music on individual tracks so they can be blended together is similar to that of adding sound effects to the picture.

Some types of programs require the use of narration to tell a story. Thus, in addition to the sound effects and music, narration might be added to other channels on the ATR. Each audio channel is identified by a number, and a log is created giving the channel number and the exact time code number of where a particular sound occurs. The sound mixer uses this log to mix all the audio elements.

In addition to the standard music, narration,

and sound effects, the director also might want to use a process called automatic dialogue replacement (ADR). ADR is used if there are problems with the original dialogue. In the ADR process, the actor or actress is brought into a sound studio and listens to the problem dialogue through headphones while the picture is projected on a screen. After several rehearsals to check for lip synchronization, desired reading, and voice inflection, the new dialogue is recorded on a blank sound track of a multi-track ATR running in synchronization with the picture.

When the director is satisfied with the best take, the rerecorded dialogue is mixed with the main production sound track in place of the bad sound. Later, these new lines will be mixed with ambient sounds, music, and sound effects in the sweetening process. Generally, the old and new dialogue are blended so well on a new composite sound track that the audience cannot tell the difference between original and replacement lines.

Once this new sound track has been created, it is married to the picture by rerecording it on the master videotape. The master on-line edited picture and the new composite sound track represent the final product.

DUPLICATION AND SYNDICATION

Once the program has been completed, it is important to make a backup, or protection copy, of the master as soon as possible. This is the best insurance you have that nothing will happen to the finished videotape.

The backup tapes, or submasters, can be used to generate additional copies for distribution. Should any of the submasters become damaged, another submaster can then be made from the master tape. Some clients prefer that distribution copies be made from the master, but it is still important to have backup copies of the master in case it becomes damaged in the duplication process.

The quality of the submaster depends on the quality of the original material and how carefully the video and audio signals were processed during editing.

Preparing for Editing 4

The biggest obstacle to a productive editing session is lack of preparation. To help you understand more about the requirements for editing, the material in this chapter relates to videotape production, which ultimately affects post-production and your job as editor. In some cases, the editor is asked to contribute his or her input prior to the start of production, and the information presented here will help you provide the kind of input the producer will find most useful.

The videotape editing room is like the small end of a funnel. Here the efforts of the entire cast and crew are represented by a few rolls of videotape, and it is the editor's job to mold this raw material into a professional product. If all the parts required to build the product are not provided prior to the editing session, the product cannot be completed. It is your responsibility as editor to make sure the producer, director, and production crew know what those parts are before they begin the production process.

PRE-PRODUCTION AND POST-PRODUCTION MEETINGS

Whenever possible, the editor should set up pre-production and post-production meetings to iron out any problems that might arise during taping or editing. The director should have some idea of what types of optical or digital effects he or she wants to use and discuss these with the editor or technical supervisor before going ahead with them. Sometimes an effect might have to be modified or replaced with something else because of the limitations of the editing equipment. In addition, by discussing these things beforehand, the editor might be able to suggest a better or less expensive way to accomplish what the director wants.

It is also important to establish the type of time code that will be used throughout the project and to make sure everyone involved with the production and post-production process knows what it is. It is most embarrassing and troublesome to find out that some of your tapes have one type of time code and the rest the other. The two types of time code—non–drop frame and drop frame—are discussed in Chapter 5. Both types are used in the television industry today, although non–drop frame is being phased out.

POST-PRODUCTION BUDGETS

Generating a budget takes time, and a wealth of experience is gained by working in some part of the television industry that deals directly or indirectly with most phases of production and post-production. Budgets are not the editor's responsibility. The associate producer or unit manager usually prepares the budget for a project. The only time the editor might get involved is at the beginning of a series, during planning for a special project, or when estimating the amount of time and type of equipment required to produce a special effect.

The following information is applicable to all types of post-production budgets, whether they are for industrial, educational, or entertainment productions. I deal with the most important items the associate producer or unit manager must take into account when estimating cost. (Figure 4.1 is an example of a typical post-production budget form.) Because of inflation and cost increases and variations, estimates are given only in percentages.

It is important to understand the terms *above the line* and *below the line* when dealing with budgets. (The line is an imaginary reference point on which all budgeting is based.)

Above the line refers to the production staff, including the executive producer, director, producer, associate producer, office staff, office rent, cast, extras, stand-ins, production assistants, and anyone else who works with the people in front of the camera. In addition, production costs and expenses such as show insurance, license fees, and talent fees are part of

SAMPLE BUDGET FOR POST—PRODUCTION

CLIENT ____________ PHONE ________ DATE ________

ADDRESS ________________ CONTACT ________

SERVICES REQUIRED	HOURLY RATE	X HRS OF USE	= COST
PREPARATION			
Window dubs			
Format conversion			
Time coding master tapes			
Film to tape transfer			
Audio to tape transfer			
Offline editing			
Edit black stock			
Work print editing			
Edit list cleanup			
Online editing			
Digital effects			
Color correction			
Color camera			
Title camera			
Electronic graphics			
SOUND SERVICES			
Laydown			
Spotting			
Laugh or applause			
Foley sound effects			
Announce booth			
Sweetening (dubbing)			
Audio stock rental			
Audio stock purchase			
Layback to master			
ADDITIONAL COSTS			
Additional 3/4 stock			
Additional 1-inch stock			
Overtime costs/person			
Additional personnel			
DUPLICATION			
One inch			
3/4 inch			
VHS			
Beta			
Other ________			

Subtotal of post-production costs: ________

Sales tax: ________

Total estimated post-production budget: ________

FIGURE 4.1
A sample post-production budget form.

the above the line budget. Miscellaneous items, legal and accounting services, travel expenses, shipping, and messenger services also are above the line expenses. Finally, music is sometimes included in the above the line budget. Music expenses include the musical director, the conductor, the orchestra, a rehearsal pianist, music clearance and instrument cartage.

Below the line is everything else. This includes the technical staff, wardrobe, studio, associate directors, stage managers, and all the personnel hired directly by the studio. Also included in this figure are the editing, raw stock, cue card service, and all incidental costs relating to the show. Telecine, and audio mixing (sweetening) are all below the line expenses.

The sample budget in Figure 4.1 is designed to cover all the areas of post-production that are affected by the editing budget. In the following paragraphs, I discuss only those aspects of the below the line budget that deal with post-production, especially editing.

Estimating Editing Time

Experience is the key to estimating editing time. If a unit manager has worked with the same director or editor before, he or she probably has some idea of how much time the editing process will take. In addition, different editing facilities operate under different regulations, often depending on whether or not they are union shops, and these regulations can affect the amount of editing time required for a specific project.

Another factor to consider is whether the show will be shot film-style, with one camera, one tape machine, and many different angles of the same scene, or with multiple cameras.

Let's take a brief look at multiple-camera production, which is normally conducted in one of two ways. In the first method, several cameras are connected to a video switcher, which is capable of combining or individually switching camera signals. In the early 1970s, the common practice was to connect five or more cameras to one switcher and have the technical director, upon command of the director, switch cameras.

The other method, used more commonly in the 1980s, is to record the output of each camera directly on its own VTR. In addition, a separate VTR might contain a switched feed, which records the combined output of the video switcher. Typically, a switched feed is used to reduce the number of electronic edits required. The final product is a combination of a switched feed, which is generally used as much as possible, and pieces from the isolated camera tapes.

Multiple camera productions that use a video switcher require less editing time than those in which the output of each camera is fed directly into a separate VTR. In the latter case, the editor will have to make many more cuts than if the show was completely or partially switched live. Similarly, single camera film-style productions might take up to three times more editing time than if the production was shot with multiple cameras and at least partially switched in the studio.

The most important thing is to talk to as many people as possible when estimating editing time. The unit manager should especially ask the director and editor how long they think it will take to edit the show, combine these figures with his or her own estimates, then add a contingency figure of 20 percent. This figure will allow for extra time needed to make producer, director, or network changes and should allow for unexpected problems.

After the unit manager has a good idea of how long the editing will take, he or she should contact the sales and scheduling departments at the editing facility that will be used. The unit manager can use the facility's rate cards to determine how much equipment and labor will cost, but often a facility will offer a special package deal that can meet the manager's needs more cheaply. For instance, a facility might offer a special deal at certain times of the year when it is not as busy as usual. Carefully examining the production's needs and the facility's options can give the unit manager a good idea of what the costs will be.

Contingency Factors

Generally speaking, a unit manager should add a contingency factor of 5 percent of the above the line budget to allow for unexpected expenses. Because of the complexity and time constraints of post-production, a contingency factor of 10 percent of the below the line expenses should be used. Although figuring in these contingency factors helps prevent budget overruns, it is important for the unit manager to estimate production and post-production costs as accurately as possible.

Insurance

Many types of insurance are available to cover potential problems during production and post-production. One of the more important types is called errors and omissions. This type of insurance will protect against charges of plagiarism and should be purchased for at least a year and longer if there is a chance the show will be rerun.

Another important type of coverage is print or tape insurance, which covers damage to the original material during the production or post-production process. This type of coverage is also referred to as accidental erasure insurance.

Workmen's compensation insurance is a must to cover people hired directly for a production. If production or post-production work is done at an outside facility, the facility's staff is covered by its own insurance.

OUTLINE OF PRODUCTION REQUIREMENTS

If the following suggestions are implemented during production, they will reduce the possibility of errors or extra expenditures during post-production. Some of these points are described in detail in the following sections.

1. Take notes during production and record them in a special log.
2. Use slates to identify material.
3. Before taping begins, specify which type of time code will be used for the entire production.
4. Record program audio on channel 1 and optional channel 2, with time code on channel 3. The time code should be recorded at what is referred to as zero level. This format is used on most major computer assisted editing systems. Newer VTRs by Sony and Ampex require a 0 db time code level, but pre-1980 model VTRs generally require a time code level of 3 db over zero level (+3 db).
5. If time code characters are inserted into the picture on work prints, follow these general guidelines:
   - Place the inserted characters in a black box if the character inserter has this capability. This makes it easier to read the numbers.
   - Don't make the characters too large. Legibility is the prime goal. It is preferable to position the characters in the lower left-hand corner of the picture to avoid covering up too much of the video. Also, don't make the characters too wide; leave room for additional time code characters that might be added during editing.
6. Whenever possible, record a tape leader at the beginning of each roll of videotape. Generally, this consists of 30 to 60 sec-

onds of color bars, time code on channel 3 (the address track), and a reference audio tone on both audio channels.

7. Check the video, audio, and time code levels periodically during production to prevent distortion or poor recordings.
8. Identify all tapes as clearly as possible. Be sure that all work tapes have the same identification as their corresponding masters. The most important pieces of information are the show title, episode title (if applicable), reel number, date recorded, whether the tape is a master or isolated feed, and what type of code is on the tape. Also include any technical information about the tape if there is a problem and promptly notify the proper people of any technical problems.
9. After recording, be sure to rewind all tapes back to the beginning. This is especially important when using videocassettes. Also be sure to check the tape label against the label on the box.
10. Once you have determined the kind and number of special effects that will be added to the program, find out whether the editing facility you plan to use is equipped to generate them. Also be sure that your budget can handle the cost of using special effects generators. Each device may cost several hundred dollars an hour over and above the normal editing rates.
11. Familiarize yourself with the editing system you plan to use. Also choose your videotape stock carefully based on performance and the editing system's demands.
12. If you are going to do an off-line edit prior to building a master tape, be sure you have an adequate supply of edit black cassettes to build your program. (See Chapter 7 for a detailed discussion of edit black.)

PRODUCTION NOTES

For production and editing to go smoothly, it is imperative to take good notes. Without them costs can far exceed the estimated budget. An example of poor note taking occurred during a television special I worked on several years ago. Because of budget constraints, an unqualified person was hired as production assistant. When recording the time code numbers, the production assistant read the wrong clock. Thus when we began editing according to the numbers in the written log, the computer could not find the time code numbers we entered. After much head scratching, we discovered the problem. By comparing the production assistant's time code notes with those compiled by the tape operator during production, I came up with a modified time code list that enabled us to begin editing. Even though we solved this problem, editing was delayed more than a day and a half.

Another aspect of note taking might be referred to as pre–post-production. Get work copies of a program on three-quarter-inch, Beta, or VHS tape with time code numbers burned into the picture, then view these in your office or home and prepare editing notes while you are watching them. This advance preparation can save a lot of time, and consequently money, when you get to the editing room.

RECORDING LOGS

One way to keep track of production material is to use a recording, or tape, log (Figure 4.2). The name of the show (if a special) or the name of the series and the episode number should appear at the top of each page. The date is also important if more than one segment was shot on any given day.

The first column in Figure 4.2 indicates the tape reel number. The second column defines the scene and take. Columns 3 through 10 indicate the approximate time code start and stop points. Only the hours, minutes, and seconds are written down, since this information is used just to locate the slate (discussed below) that identifies the corresponding scene. The frames columns (columns 6 and 10) are used to identify specific frames for special effects and other special uses.

The comment space might be used to indicate whether a take is good or bad. The comment column also might be used to record any information that might be helpful during the editing process. To avoid excessive paperwork, each page should have enough space for at least a dozen takes.

SLATES

A slate is merely a convenient way to locate material on the tape. Since videotape is no longer cut apart scene by scene, all the takes are contained on large reels. For example, the reel for a comedy show might contain more than 100 takes of short jokes that will appear

Production log

PAGE 1 OF 6
DATE MARCH 1, 1988

CLIENT: INDUSTRIAL PRODUCTS
ADDRESS:
TITLE: NEW PRODUCT DEMO

PRODUCER: ROY BEAL
DIRECTOR: MARY SMITH
PHONE: EXT 4581

REEL	SLATE	HOURS	MIN	SEC		HOURS	MIN	SEC	✓	COMMENTS
1	1-1	08	45	22		08	45	59		Opening Logo
1	1-2	08	47	01		08	47	49	✓	Logo with push in-
1	2-1	08	53	10		08	54	20	✓	Opening remarks
1	2-2	08	55	27		08	56	33	✓	pickup - top of page 2
1	2-3	08	57	31		08	58	58	✓	pick up top of page 4
1	3-1	09	00	20		09	02	15		truck shot around table
1	3-2	09	15	17		09	17	20	✓	" " better move
2	6-1	10	54	16		10	54	57	✓	C.U. circuit board
2	6-2	11	01	01		11	01	41	✓	tighter shot - hold
2	7-1	11	22	00		11	23	02	✓	product checking
2	8-1	11	28	16		11	29	43		C.U. Bill checking
2	8-2	11	33	03		11	34	02	✓	" " no smile
3	4-1	11	56	07		11	56	21	✓	inserting cotter pin
3	5-1	12	06	11		12	06	51		installing chip
3	5-2	12	07	21		12	07	41	✓	" "
3	6-1	12	15	15		12	15	35	✓	installing power cord
3	7-1	12	17	01		12	17	43	✓	connecting speaker leads
3	8-1	12	18	18		12	18	33	✓	seating PC Board
3	9-1	12	22	06		12	23	03		connecting disk drive
3	9-2	12	24	30		12	25	00	✓	" " "

FIGURE 4.2
A typical log used during production that might assist the editor during post-production.

in the final product. You can imagine how difficult it would be to locate anything on this reel without accurate and concise notes. The slate, along with time code information, helps you locate the correct take quickly and easily.

Each take, segment, scene, or sequence should be identified by a slate. The rule of thumb is to use a slate each time the VTR starts to record. Sometimes, if an actor flubs a line at the beginning of a take, the stage manager will put the slate in front of the camera to identify the next take without stopping the VTR.

Slates can be made of almost any type of material. Commercially available slates are usually made of a light wood such as balsa or Masonite and are painted a dull black so as not to reflect glare from the lights. Dull white lettering is used to indicate areas in which you can write frequently changing information. Most slates have a blackboard-type surface so that you can write on them with chalk, but any form of scene identification is valuable no matter how crude it might appear.

As soon as the stage manager finishes slating a scene, he or she is usually busy writing the next scene and take number on the slate so that it will be ready when needed. As the slate is being recorded, the script person logs this information in the editing script.

It should be noted that the recording log, or tape log, contains time code information and comments. The editing script is a marked copy of the program script, with additional comments from the production staff and sometimes even time code notes relating to a specific scene or take that the director may want to use. Vertical lines may be drawn on the script to indicate the beginning and end of a scene or take. Each vertical line may be headed by a scene or take number.

Slates should be printed legibly and with accurate information that includes the show title, episode or production number, director's name, date, scene or sequence title, take number, and page number of the script (optional). There are many methods of identifying a scene. Motion pictures use numbers for master shots and numbers followed by a letter for pickups or inserts (for example, scene 112A). This works well for most applications, but it is possible to get confused if a scene has many inserts.

To my mind the simplest method of identifying material is to use a consecutive numbering system. You merely start with take 1, regardless of where that scene goes in the script, and number consecutively thereafter. When you get a good take, you just record that number in the script. This helps eliminate confusion since no number is used more than once and no letters are used at all.

As mentioned before, the editing script contains the take numbers, specific information concerning the length of each take, whether it is good or bad, and other pertinent information needed for editing. Even with this written record, the editor still relies heavily on the slates. Thus it is important that the slate is held steady, the camera person focuses on the slate, and the information is complete and legible.

RECORDING TIME CODE

The primary cause of time code problems is carelessness. To edit successfully by computer, the time code on each roll of videotape must be ascending and consecutive—that is, the code must start with a low number and increase throughout the roll without backtracking.

Time code is applied to videotape before, during, or after material is recorded by means of a time code generator. One way to prevent discrepancies in time code is to apply the time code during production. To do this, the tape operator connects the time code generator to the tape recorder so that when he or she starts to record, the time code is automatically applied. When the recorder stops, the time code generator also stops.

Using the time of day is another way to avoid time code problems. The time code generator is preset with the time of day, which means that the time code clock runs continuously, even when the recorder stops. Even though there might be small gaps of time between takes, this is not a problem during editing because the time code numbers are still ascending and consecutive.

The only problem with this method is that confusion can arise if the director is recording on consecutive days. For example, imagine that the director starts shooting at nine o'clock on the first morning. The first time code would read somewhere around 09:00:00:00. The director then finishes shooting at about three o'clock in the afternoon, and the ending time code reads 15:00:00:00, which corresponds to 3:00 p.m. on the 24-hour clock. So far so good.

The second day the director again decides to start shooting at nine o'clock in the morning and, to cut costs, decides to use the balance of the roll of videotape started the day before. This is where the problem occurs. Because both days' shooting began at nine o'clock, this time code will appear twice on the same reel. If the editor asks the computer to look for this time code during editing, it will not know which day's code to access.

There are two ways to get around this problem. First, the recording log kept in the box with the tape could notify the playback tape operator of the conflicting time code. If the operator is alert, he or she will manually park the VTR in the right code area so the edit can be made.

The second and preferred way to avoid this problem is for the director to make sure each day's recording begins on a fresh roll of tape. If the director must record on a previously used roll of tape, someone should manually reset the time code generator to any time greater than the last time code number recorded on the tape.

In this case, the time code generator is no longer reading time of day, but as long as proper notes are taken during production, this method should pose no problem during editing.

RECORDING LEADERS

It is imperative to record a sufficient leader ahead of the action cue if a computer assisted editing system will be used in post-production. The purpose of this is to allow the recording equipment to reach a stabilized speed. The tape operator generally should allow about 15 seconds from the time the record button is pressed until the first action cue is given by the director.

This 15-second stabilizing period can be used to slate scenes and to focus and frame the cameras. Once the VTR has started to record and as long as it continues to record without stopping, another scene can be slated without waiting for a 15-second leader. Small gaps of program material and time code are acceptable. But, it is wise to avoid large gaps of blank tape between takes because these might confuse some editing systems. To be safe, there should be no more than a five-second time code gap between takes. A detailed study of time code and its relationship to editing appears in Chapter 5.

The leader is used during editing to allow the VCRs to cue up ahead of the desired point. Although newer VTRs and VCRs generally use a 5-second leader ahead of an edit, older VCRs might require a 10- to 12-second leader. The absence of a leader can be very frustrating for the editor.

LABELING TAPES

One of the more frustrating problems in editing is the lack of proper identification on a roll of videotape. An incredible amount of time might be wasted during an editing session looking for a tape that was incorrectly labeled. Editing room charges do not stop just because you cannot find a roll of tape.

Following is a list of items that should be included on both the inside reel label and the outside box label of each tape. If you use preprinted labels, you might be able to design the label as a checklist so that you just have to check off boxes for each reel. Always write or print clearly and make sure the reel label and the box label contain the same information.

1. Name of the production company or producer
2. Working title of the show
3. Part number if the material is longer than one roll of tape
4. Running time of the reel or time code start and stop times
5. Tape generation (Is it original production material? Is it an edited master? Is it a copy of the edited master, which might be referred to as a dupe, submaster, B roll, protection master, or dub? Is the tape a direct transfer from film?)
6. Duplicate (Dupe) number if it is a copy of an edited tape
7. Reel number if it is to be used in editing
8. Date the tape was recorded (especially important if the production was recorded over a period of several days)
9. VTR number (in case there is a technical problem with the tape) and tape operator's name or initials
10. Type of audio used on the tape
11. Tape format (One-inch B and C format tapes look the same but are incompatible, so it is important to identify the format.)
12. Number of times the tape has been played, especially if the tape is used in syndication (This should be updated each time the tape is played so that there is an accurate record of tape usage and wear.)
13. Tape signal or standards format (This indicates whether the tape was recorded in the NTSC, PAL, or SECAM format, all of which are incompatible. See Chapter 5 for a discussion of these formats.)

Although this might seem like a lot of information, you can probably fit it on a 1-inch by 6-inch label if it is properly laid out. In some cases, you might require the use of more than one label to list all the information needed for identification.

BOOKING EDITING TIME

Failing to reserve the proper amount of editing time can cause severe time and financial problems. It is prudent to reserve anywhere from 30 to 50 percent more time than you think you will need. Many large post-production facilities charge only for the time used and not for the time booked, but it is wise to clarify this when you book the time so that you will not be charged for unused time or be surprised when you get the bill.

When booking time, it is also important to reserve any special equipment you will require. If you do not reserve this equipment in advance, you might find yourself waiting while the meter is running until the equipment you need becomes available.

Last, if you must cancel a scheduled edit session some facilities may charge a percentage of the booked time (even up to 100%), if adequate advance notice is not given in accordance with the terms and conditions set forth by the post-production facility.

CHECKING THE EQUIPMENT

Too many editors take post-production equipment for granted. That is, they assume that they know how to operate it and that when they walk into the editing room every piece of equipment has been calibrated and is working properly. That is not always true. Just as a pilot checks an airplane before taking off, an editor should check his or her equipment before starting to edit.

For example, suppose the last person using the room set up the monitors for a special effect and did not restore them to their normal state. If you assume they are set up correctly and do not bother to check, the end product may not be acceptable. It is imperative for the editor or the assistant editor to check this equipment before proceeding.

VIDEOTAPE STOCK

Many brands of videotape are used by the television industry. Some of the more popular are 3M, Sony, Fuji, Ampex, and Kodak. Generally, any of these brands works well for any application, although some facilities might prefer one brand over another.

There are two schools of thought regarding the purchase of videotape stock. One says you should buy the tape from the facility where you plan to edit. The other says you should purchase the tape directly from the manufacturer. There are a number of advantages to buying the tape from the facility. The facility guarantees that the tape has been tested and evaluated, and if the tape is defective, the facility will not only replace the stock but also redo the program material at no charge. For this guarantee, you pay a higher price for the stock.

Buying the tape directly from the manufacturer or its representative results in considerable savings. The videotape is almost always acceptable without testing and evaluation, but there is always a chance that the tape will have some defect that renders it unsuitable for your needs. In that case, you will have to pay to redo the program because the facility is not responsible for the defective stock.

Based on my experience, I strongly recommend that videotape stock be purchased from the production or post-production facility. Buying the tape stock from the facility ensures that if a problem arises—for example, if the recording is technically unacceptable—that cost will not come out of your budget.

EDITING REQUIREMENTS

Make sure all the material you need for a project is available before you start to edit. Here are some of the things you should do before you step into the editing room:

1. Transfer motion picture slides or film in any format, from 8mm to 35mm, to videotape. (Chapter 11 outlines the way motion picture film is transferred to videotape.) Reserving a telecine machine to use for this transfer during editing is expensive and impractical.
2. Have window dubs made of all the material you plan to use in the program. This includes telecine footage from film, stock footage, and any material photographed with a live television camera.
3. If some of the videotape material is in a different format than most of your production material, be sure to convert the odd pieces to the same format and to make window dubs of this material.
4. Prepare your edit black (described in Chapter 7) ahead of time for both off-line editing and on-line editing.
5. Transfer to your tape format any "wild" audio or announce lines that will be required during editing.
6. Make a backup, or B roll, of material that will be used for optical effects.

CONCLUSION

Not only is the editor responsible for the creative aspect of the product, but he or she must also establish communication with the production and post-production staff. The most important rule is when in doubt, ask. This will prevent errors, speed up production, and keep costs (and tempers) down. It is also a good idea to make sure everything is in order prior to the first day's shooting. Keeping the lines of communication open and keeping an eye on the production process will reduce the number of problems that will arise in post-production.

All About Time Code

5

Time code is an electronic method of identifying television picture frames that is similar in concept to film edge numbers. The most popular use of time code is in locating a particular frame to make an electronic edit.

Although NASA has used a form of time code to track missiles and perform various other telemetry data functions for years, it was not used by the television industry until about 1967. Before that, various methods were used to identify individual frames so that accurate edits could be made.

As early as 1958, NBC in Burbank, California, had formulated a way to edit videotape using a kinescope motion picture film work print with a custom-designed voice track that, in effect, was a very crude form of time code. Although this edit sync guide (ESG) was crude by today's standards, it was used very successfully for a number of years. (See Figure 5.1.)

In early 1963, Ampex came out with the first electronic editing system, called Editec (Figure 5.2), which recorded single-frame audio tones on the secondary, or cue, channel of two-inch videotape. Each tone, or beep, was recorded manually by the operator. The tones triggered the electronic editor that was part of the Editec system. Although the operator could reposition these tones, doing so was time-consuming and sometimes frustrating.

The primary disadvantage of Editec was that the trigger tones were relative to each edit point and had no significant relationship to sound or picture information on the rest of the tape. In other words, the beeps could not be used to locate material anywhere on the tape except at the frame indicated by the beep. The only information that could be located was in frames previously marked with a beep.

If the tape had no beeps applied to it, there would be no way to identify picture information except by playing it at normal speed. Most VTRs had linear tape timers in hours, minutes, and seconds that could be used to locate material. However, they had to be reset to zero at the beginning of the reel and would often be inaccurate by seconds or minutes due to slippage of the counter against the tape. These counters were only marginally useful. Stopping the tape on a two-inch VTR would make the picture disappear, and it was impossible to speed up or slow down the tape. However, with the advent of helical-scan VCRs, the picture could be stopped or viewed at nearly any speed within the range of the VCR's capability.

To locate individual frames on a helical-scan videotape, the VCR needed the electronic equivalent of motion picture film edge numbers.

FIGURE 5.1
The author is shown editing kinescope film using ESG code.

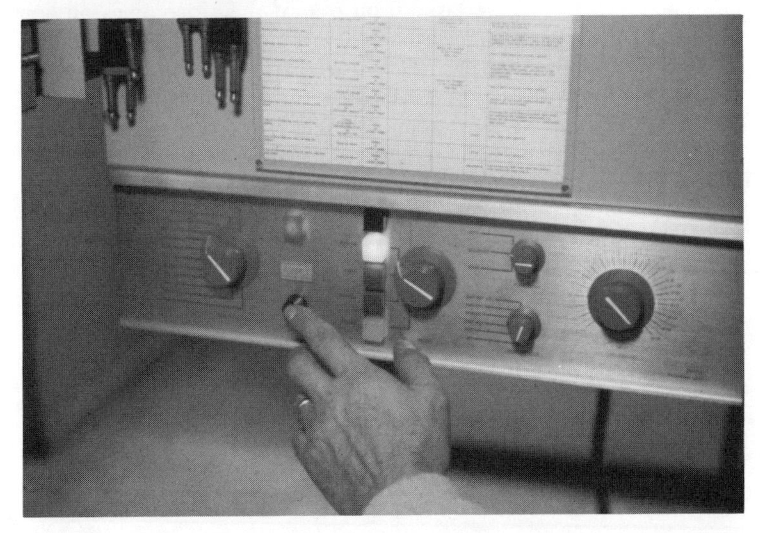

FIGURE 5.2
An early Ampex Editec electronic editor control unit.

10	23	42	12
HOURS	MINUTES	SECONDS	FRAMES

FIGURE 5.3
A typical SMPTE time code display. (Courtesy EECO)

The first manufacturer to utilize time code for editing was EECO of Santa Ana, California. For many years, time code editing was generically referred to as EECO editing because of the popularity of the EECO system. The advantage of the time code system over electronic editing systems such as Editec was that the frame numbers could be written down and edits made again with frame accuracy.

Other manufacturers introduced their own time code–based editing equipment, but most of the systems were incompatible with each other. In 1969 the Society of Motion Picture and Television Engineers (SMPTE) established a committee to develop a standard time code that would provide interchangeability. This basic time code format (Figure 5.3) also was adopted by the European Broadcast Union (EBU).

HOW TIME CODE IS FORMATTED

One of the more important functions of time code is that it identifies every video frame consecutively on the tape. This is an advantage over the use of a control track pulse, which is analogous to film sprocket holes. Control track pulses help maintain a constant tape speed but cannot be used to accurately access specific frames. Time code, however, can be used to search for specific frames, as well as synchronize one or more tapes during the editing process.

Time code information is recorded, usually by a time code generator (Figure 5.4), in a binary fashion on one of the videotape's audio channels. In the early days of videotape editing, the track used to record the time information was called the cue channel. Much later, a third audio channel called the address track was added to record time code information. This prevents any interference, or cross talk, with either of the two primary audio channels.

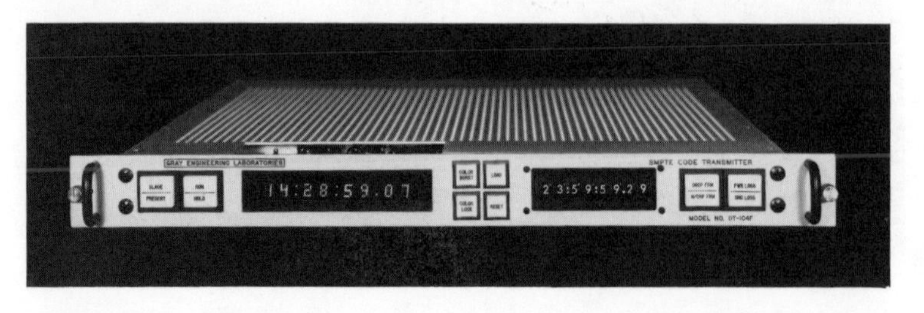

FIGURE 5.4
An example of a time code generator. (Courtesy Gray Engineering Labs)

TYPES OF TIME CODE

There are two forms of SMPTE time code—longitudinal time code and vertical interval time code (VITC). Longitudinal time code is recorded by conventional means on an address track and has been the industry standard since 1967. VITC, which has been around since about 1980, is not as widely used because of the hardware needed to record it and display it.

Because longitudinal time code is recorded on an audio track (Figure 5.5), it can be read by time code readers at up to 60 times play speed in either direction. To read longitudinal time code reliably at very high search speeds, the VTR or VCR must be equipped with wideband amplifiers and reproduction heads. This is because time code frequency increases with the speed of the tape to a point where a high-frequency roll-off similar to that found in high-fidelity amplifiers might occur in the VTR's or VCR's longitudinal time code reproduction system. If you do not use special wideband audio equipment during editing or high-speed searching, the result is an erratic time code signal that can cause frequent edit aborts or failure of the system to find a desired time code number.

At low tape speeds, the time code frequency decreases along with the amplitude, so the VTR or VCR must have good low-frequency response and might require high-gain amplifiers to detect the time code accurately. At speeds below about six frames per second, the time code level drops below the ambient noise level and might disappear completely. This is due to the fact that the tape is moving so slowly across the reproduce heads that little or no signal voltage is induced in the head. At such times, many time code readers automatically switch to counting control track pulses, synthesizing time code to maintain continuity until the reader detects a usable code. This type of control track counting is not accurate and might cause editing errors.

VITC is recorded as part of the vertical interval portion of the video signal and is detected or decoded by the video head, or scanner. It solves most of the reading problems associated with longitudinal time code and is the most accurate, repeatable, and reliable way to edit videotape. Unfortunately, the cost of the hardware to detect and display VITC is relatively expensive compared to the cost of longitudinal equipment. VITC has the ability to detect time code accurately from still frame to about 45 times play speed in either direction.

FRAME IDENTIFICATION

Longitudinal time code is an electronic signal in the audible range that changes from one voltage to another, generating a group of pulses that, when correctly decoded, are displayed as time code numbers. In the National Television Standards Committee (NTSC) standard, which is used in all of North America, parts of South America, Japan, South Korea, and several other countries, every 30 frames, or one second, is sliced into 2,400 equal parts. The rest of the world uses two other television formats, PAL and SECAM, which operate at a rate of 25 frames per second and slice each second into 2,000 parts.

Each slice of time code is called a binary bit. In any of the standards, there are 80 bits per television frame (2,400 divided by 30 or 2,000 divided by 25). A bit is the smallest unit of data used by a computer and exists in one of two states, either as a 1 or a 0, which might also be interpreted as being on or off. Of the 80 bits in each frame of time code, not all are assigned; generally 5 bits are unused.

The bits in each frame (Figure 5.6) provide three types of information. The first is the hour, minute, second, and frame count. The second is eight optional four-digit characters reserved for other data or for control purposes. The third is a special sync word that indicates the end of each frame and the direction the tape is moving. Of the 80 bits per frame, 32 are reserved for time code, 32 hold user-designated information (these are called user bits), and the remaining 16 are used for the sync word. Of these latter bits, bit 10 is used when the time code is in drop frame mode (discussed later in this chapter), and bit 11 is the color frame bit. Figure 5.7 shows the drop frame indicator in one frame of time code, while Figure 5.8 shows the color frame bit in a frame of code.

User Bits

These 32 bits of special information are interspersed throughout each 80-bit frame of time code in eight groups of 4 bits each. (Figure 5.9 shows the user bit group in one frame of time code.) They are available for any purpose the user wishes. To take advantage of user bits, you must use equipment designed for this pur-

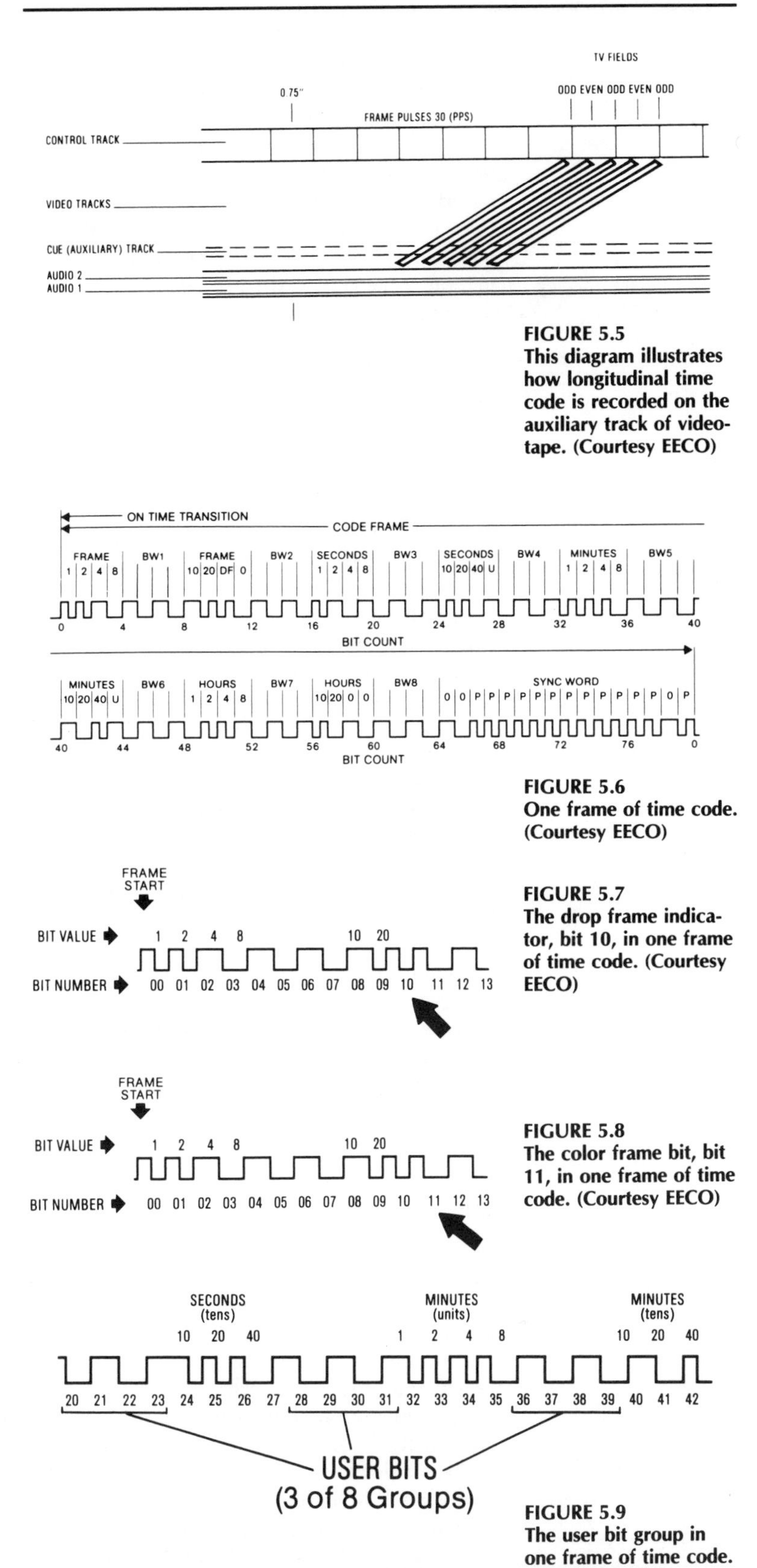

FIGURE 5.5
This diagram illustrates how longitudinal time code is recorded on the auxiliary track of videotape. (Courtesy EECO)

FIGURE 5.6
One frame of time code. (Courtesy EECO)

FIGURE 5.7
The drop frame indicator, bit 10, in one frame of time code. (Courtesy EECO)

FIGURE 5.8
The color frame bit, bit 11, in one frame of time code. (Courtesy EECO)

FIGURE 5.9
The user bit group in one frame of time code. (Courtesy EECO)

pose. Until recently, user bits were not used widely because of a lack of standardized equipment to generate and display this type of data.

User bit data must be recorded at the same time the time code is recorded and cannot be added once the time code has been put on the tape. It is up to the tape operator or someone on the production staff to provide the information to be recorded in the user bit area of the time code.

User bits can be helpful in identifying reel numbers or providing other information, but they must be used consistently. It is of no use to have some reels with user bit information and others without it.

Some editing systems rely on user bits for reel identification. Instead of having the editor type in the proper reel number each time the tape is changed, the system selects the reel number from the user bit information and automatically enters it in the computer's memory.

One of the more common methods of editing musical numbers is to put time code on an audio playback device used during production and transfer that information in user bits to the videotape. If the same section of audio is played back repeatedly, the time code is also copied on the production master videotape the same number of times. A computer assisted editing system identifies the correct take of the musical number by detecting the user bit code in the time code on the audio track. As the time code of that take is repeated over and over again, it is being recorded on the production master tape along with a time of day code which is also transferred to the production master videotape on a continuous basis so that each take on the production master tape now contains two different codes, one repeating (user bits) and one in ascending order (time of day).

Color Frame Bit

Bit 11, the color frame bit, tells the time code reader whether or not the color frame identification has been applied intentionally. What is meant by color frame? In the NTSC color television standard, a complete color frame is made up of four television fields (or two picture frames) equal in time to one fifteenth of a second. In the PAL and SECAM television standards, a color frame is made up of eight fields, or one sixth of a second.

In the NTSC standard, fields 1 and 3 are defined as color frame A, while fields 2 and 4 are defined as color frame B. Color framing includes alternating positive and negative values assigned to each television frame. If two consecutive television frames are electronically spliced together, one frame must have a positive value and the next frame a negative value. Once established at the beginning of a tape, color framing maintains the same relationship correctly throughout the length of the tape. Failure to achieve proper color framing will result in a noticeable horizontal picture movement at the edit point.

Problems with color framing often arise when the editor is generating an optical effect such as a dissolve. At the point where the scene starts to dissolve out, a horizontal picture shift will occur if there is a color frame problem. The outgoing picture will move slightly to the left or right at the edit point, indicating that either the color frame is not correct or the time base corrector jumped one position away from its normal state. Assuming that the color frame is not correct, most editing software gives the operator the opportunity to adjust the color frame by changing its relative phase with respect to the edit by 180 degrees. When the edit is remade, the color frame software tells the VTR or VCR to correct its color phase electronically, putting the video image back in "electronic time," in effect, repositioning the video signal at the edit point thus eliminating the horizontal picture shift. After the edit is remade and recorded, it is replayed to determine whether the picture shift was eliminated. If it was not, the tape operator should adjust the time base corrector on the source playback VTR or VCR to try to correct this problem.

Color frame problems also arise when an editor is electronically splicing together or extending a scene. The splice point should not be visible when it is viewed on the monitor. If the picture shifts left or right as the tape is played, the color frame must be changed and the edit remade. As a rule, once color framing has been established at the beginning of the tape, color frame errors will not occur. This type of error may occur when editing into a previously edited master tape, so the precautions just outlined should be followed.

TIME CODE VERSUS CONTROL TRACK EDITING

One of the reasons SMPTE time code was developed was to ensure frame accuracy and repeatability during editing and post-production.

This was one of the shortcomings of control track editing, which is not frame accurate because there is no way to index tape frames on a repeatable basis. The only reference is the control track pulse, which is applied electronically at the time the videotape is recorded. There is one pulse for every frame, but they are all identical, so there is no practical way to tell them apart.

Why would someone buy an editing system that is not frame accurate? The primary reason is cost. Control track editors are less expensive because they do not include the sophisticated hardware needed to read and generate time code.

Control track editors (Figure 5.10) make edits by counting control track pulses rather than reading time code. On all videotapes, regardless of size or format, there is one control track pulse per television frame. The hardware and software designed into the system count these pulses and use them to determine where to start previewing or recording an edit.

When an edit point is selected, most control track systems subtract 150 frames (or five seconds) from the indicated edit point, then rewind and park each VCR involved in the edit at this point. Once the VCRs are parked, all the machines go into play and the edit is made 150 frames later. Chances are that the point at which the edit actually occurs on the tape is not the exact point originally selected during the preview. This is because there is a certain amount of slippage during the time the VCRs are reversing themselves. Thus it is virtually impossible to get exact repeatability of an edit.

Most editing systems define the amount of preroll in seconds rather than frames. Although most systems have a fixed preroll of 150 frames, some systems allow you to reduce the preroll to only 2 seconds or increase it to 30 seconds. The shorter the preroll, the less chance there is for slippage. Although this might improve frame accuracy, it can cause a bad edit because the VCRs involved in the edit might not be stabilized and a disturbance at the edit point may result.

To maintain some sort of control over frame accuracy, the edit point should be readjusted after every two previews. Many people become familiar with how much their editing system slips and will deliberately misadjust the edit point based on the slippage. This generally improves editing accuracy.

Many editors who use control track editing systems build edited work tapes by using a

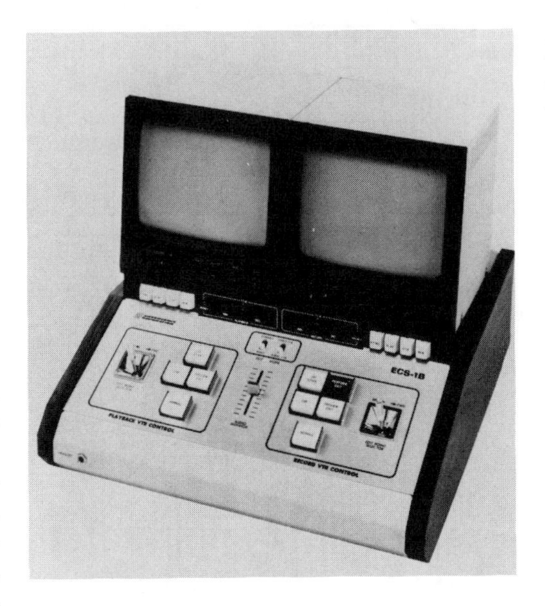

FIGURE 5.10
The ECS-1B control track videotape editor. (Courtesy Convergence Corporation)

window dub with time code information imprinted on the picture. By writing down each in and out edit number displayed at the edit point on the monitor, they can generate a frame-accurate edit list on paper. (See Figure 5.11.)

Although this process is somewhat tedious and time-consuming, the time code numbers can be adjusted to compensate for edits that the system did not make as desired. Thus, if the edit was not made at the exact frame desired, the user can add or subtract frames to the number displayed in the time code window at the edit point, store this information on a floppy disk, and then use the disk in a time code–based editing system.

Control track editing systems are adequate for some applications. These include the rough assembly of sequences and the building of a presentation reel, both of which generally do not require frame accuracy. If frame accuracy is required, a time code–based system is essential.

REQUIREMENTS FOR TIME CODE EDITING

To read longitudinal time code accurately on videotape, the editing system must include the following basic hardware: a time code reader (Figure 5.12) for each VTR, a solid reference signal such as video black provided by an internal or external black generator, and a synchronizing source to maintain stability within

FIGURE 5.11
A frame-accurate handwritten edit list.

EDITING LOG

PAGE 1 OF 12
DATE 3-29-88

CLIENT: INDUSTRIAL PRODUCTS
ADDRESS:
TITLE: SALES PRESENTATION # 2

PRODUCER: ROY BEAL
DIRECTOR: MARY SMITH
PHONE: EXT 2341

REEL	SLATE	HOURS	MIN	SEC	FRAMES	HOURS	MIN	SEC	FRAMES	COMMENTS
										BARS/TONE FOR 60 SEC.
										SLATE 10 SEC.
										BLACK 20 SEC.
01	1-6	08	43	17	23	08	43	41	21	30 FR. FADE-IN A/V
17	1A-1	14	10	16	01	14	10	18	01	Video insert - BACK IN
										FROM END OF PREV. EDIT
01	2-2	08	51	07	11	08	52	00	03	DISS. FROM INSERT TO V.P.
01	3-1	09	04	23	00	09	04	49	21	A/V "FROM THE.....LATER"
02	4-1	09	56	59	14	09	57	12	12	VID. INSERT TO COVER EDIT
02	5-3	10	02	02	02	10	02	08	11	V-ONLY PLATING TANKS
02	6-1	10	02	40	00	10	02	41	15	V-ONLY CU HANDS
02	7-2	10	03	03	21	10	03	07	19	V-ONLY LS TEST SETUP
02	8-2	10	04	14	10	10	04	15	16	V-ONLY CU DIAL FACE
03	9-1	10	11	00	06	10	11	01	29	V-ONLY PAN TO PC BOARD
03	10-1	10	14	57	21	10	15	00	21	V-ONLY TRUCK PULLS AWAY
										DISS 30 FR TO HIWAY.
06	11-2	12	12	14	21	12	12	44	07	A-ONLY NARRATION TO
										COVER LAST 6 V-EDITS
07	12-4	09	11	07	15	09	11	31	17	A/V MR. CLARK 40 FR
										FADE TO BLK
17	13-1	14	03	22	21	14	03	30	00	FADE-IN TO CU TRUCK TIRE
03	14-1	10	36	17	24	10	36	21	29	CU FUEL PUMP
03	15-2	10	40	41	23	10	40	43	22	VALVE - USE AS VID INSERT
										OVER END OF SC 13-1

the system. If the system does not include these features, frame accuracy will be inconsistent at best.

Time code systems do cost more than control track editors, but if you want to create any optical effects, the convenience is worth the extra cost. If frame accuracy and repeatability are not considerations, then a simple control track editing system should suffice for most applications.

NON–DROP FRAME VERSUS DROP FRAME TIME CODE

When time code first appeared in 1967, it was called continuous time code and was referred to as color time. It is now called non–drop frame time code, meaning that no frames are dropped during the recording or displaying of time code data. Non–drop frame time code is recorded at the television color frequency, which makes the tape and the time code 3.6 seconds longer than each hour of clock time. This caused a lot of confusion because a one-hour show edited with non–drop frame time code as a timing reference would play back 3.6 seconds longer in real time.

To solve this problem, the SMPTE standards committee came up with a new form of time code that compensates for this 3.6-second increase. This format is called drop frame time code. The committee decided to eliminate 108 frames each hour (3.6 seconds are equal to 108 television frames) to make the time code match clock time. That translates to dropping two frames per minute, except at the 10-minute marks (that is, 10, 20, 30, and so on). The frames are dropped as the time code changes from one minute to the next. For example, two frames are dropped after 01:04:59:29, with the next time code being displayed as 01:05:00:02, skipping frames 00 and 01. Figures 5.13 and 5.14 show the non–drop frame and drop frame time code relationships to picture length.

FIGURE 5.12
An SMPTE time code reader. (Courtesy Gray Engineering Labs)

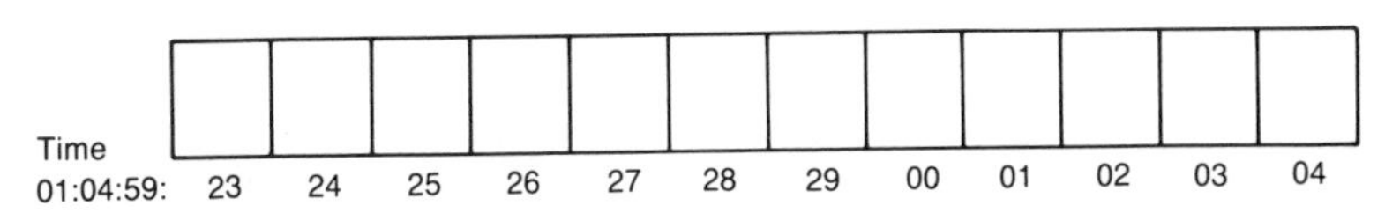

FIGURE 5.13
Non–drop frame time code relationship to picture length.

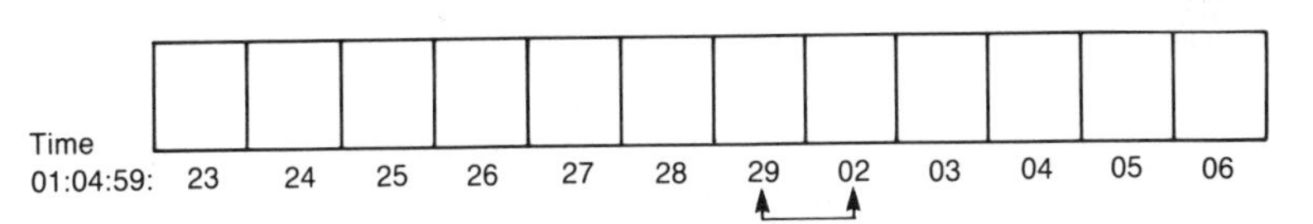

FIGURE 5.14
Drop frame time code relationship to picture length.

AVOIDING TIME CODE PROBLEMS

As mentioned in Chapter 4, one of the most frustrating problems an editor can encounter is the lack of consistency in recording time code during production. Although most videotape editing systems are capable of dealing with either drop frame or non–drop frame time code or, in some cases, both, producers frequently do not take the time to make sure that one type is used consistently on all the tapes. This is especially important when second unit production teams are shooting material on location or when prerecorded material is being supplied by an outside source. Unless the producer or someone on his or her staff maintains this consistency during production, post-production costs and time will increase because of technical problems associated with mixing time code.

Even more important is the placement of time code on the proper audio channel. As mentioned earlier, longitudinal drop frame time code should be recorded on the address track, also called channel 3 of the audio track, when using one-inch videotapes and three-quarter-inch videocassettes. If older equipment is being used, time code should be placed on channel 2. In this case, channel 1 becomes a mono audio channel to record or play back program sound, and stereo sound editing is not possible.

Another problem occurs when the work tape has time code in the window but no corresponding code on channel 2 or 3. Even though some editing systems are capable of reading control track pulses and simulating time code based on these pulses, they are not frame accurate due to mechanical slippage. A window dub should always include the corresponding time code on the address track of the videotape or on channel 2 of those older VCRs not equipped with an address track option.

Despite the fact that time code waveforms (a record of the pulse frequencies) might look normal during recording, many factors can distort the time code signal to a point where time code readers cannot reliably recognize the code's digital 1s and 0s, even at normal speed. The problems are even worse at speeds above or below normal play speed. The following suggestions will help you maintain quality and reduce or eliminate time code problems with any VTR or VCR.

The most serious problem encountered in time code editing is poor quality time code. Some of the factors that contribute to this are dirty audio heads, distortion of the time code signal through amplifiers, misaligned audio heads or tape guides, and attempts to duplicate time code without proper signal processing. Dirty heads can cause low signal levels, distortion, or no signal at all, so it is important that they be cleaned regularly. Audio head alignment should usually be checked by a maintenance engineer with a master alignment tape provided by the VTR manufacturer. Don't take it for granted that the last person who used the machine cleaned it. A few minutes of care can prevent many costly and time-consuming problems.

Time code also should be recorded at the manufacturer's recommended level. Nearly all computer assisted editing systems use channels 1 and 2 to record program audio and channel 3, also called the address track, to record time code. Current technology dictates that time code should be recorded at 0 db—that is, the time code should be recorded and played back at the 0 db reference mark on the volume unit (VU) meter. The only VTRs that use higher time code recording levels are the now obsolete two-inch quadraplex VTRs and older VCRs with only two channels of audio, such as the Sony 2850 and 2860 series.

There are several reasons why time code might not have been recorded at the recommended level. These include an electronic problem, a faulty tape, and dirt. Recording time code at levels below 0 db might impair searching, cueing, and time code readability in some systems. Other systems might accept time code levels as low as −5 db, but only if the quality of the time code is good. If consistency is maintained when recording time code, chances of encountering problems on any videotape editing system are slim.

Occasionally, audio amplifiers will become overloaded because of excessively high levels of time code. This almost always causes distortion in the time code signal. The solution is to use high-quality audio amplifiers and distribute the code without overloading the system. Finally, a time code problem can occur if the time code is fed into an automatic gain control (AGC) amplifier or through an audio limiter.

SLAVING TIME CODE

Slaving, or jam syncing, time code is a procedure that reads time code from the videotape and synchronizes a time code generator to the exact code on the tape. Once synchronized, this slaved time code is rerecorded on the videotape. The purpose of slaving is to correct poor quality or marginal time code by regenerating it as fresh time code.

Slaving may also be used with the assemble mode of electronic editing to continously add time code as you record and edit.

Care must be taken not to erase existing good time code, as doing so will affect the editing notes for the following material. For instance, if there is a gap in the time code between takes and you wish to replace the time code on the first take, it is important that you slave the time code on only that take and not let the slave time code generator continue recording over the second take. Otherwise, the new time code numbers, which will not include the original gap, will not correspond to the editing notes for the second take.

DUPLICATING TIME CODE

When copying time code from one tape to another, the safest way to prevent distortion is to duplicate it through a master slave clock or internal VTR time code generator that generates fresh time code in synchronization with the source time code. If the time code reader is equipped to regenerate time code, this too may be used to reshape the time code waveform so that it will read without distortion. It is not advisable to copy time code by any other means because the quality will not be adequate for editing.

NONSYNCHRONOUS TIME CODE

This problem manifests itself as frequent aborts of the edit cycle (either preview or record). The system will repeatedly attempt to make an edit, but as the incorrectly recorded time code approaches the edit point, the code will drift out of frame synchronization and straddle two adjacent frames. The editing system gets confused because it cannot determine which frame is the correct one. Therefore, it recues and tries to make the edit over and over again.

The problem of nonsynchronous, or drifting, time code might occur when the reference signal feeding the time code generator is interrupted. Note that time code can be recorded correctly only when the VTR is referenced to a video signal that is locked horizontally and vertically. Otherwise, the time code will not synchronize with the television picture frame.

The only way to correct this problem is to recode the section of tape that exhibits this error. The jam sync method of slaving time code maintains the same time code relationship as currently exists on the rest of the tape and preserves continuous frame accuracy.

TIME CODE NOTES AND EDIT LOGS

These notes help the editor locate scenes, identify good and bad takes, and supply the staff with information that relates to editing. Basic information should include the scene number, take number, page number in the script (optional), start time code where the slate first appears, start time code of the scene (action cue), stop time code of the scene (when "cut" is given), and comments about the scene. The editor should receive the original, and a copy should be included in the box with the videotape.

The tape and box also must be carefully labeled. See Chapter 4 for a description of the labeling process. These labels complement the log and allow the editor to do his or her job more efficiently.

RECORDING MULTIPLE FEEDS

When two or more isolated cameras are being recorded at the same time, it is imperative that the tape operator carefully check to make sure all the VTRs involved in the recording session are being fed from the same time code generator and that either drop frame or non–drop frame code is used consistently throughout the session.

ASCENDING TIME CODE

As discussed in Chapter 4, time code must always be recorded in an ascending manner. When starting to record time code at the beginning of a reel, always start with a low number and let the time increase to a higher number. If you are recording on a reel that already contains material recorded at an earlier time, make sure you reset the code generator to any number greater than the last number recorded on the reel. See Chapter 4 for a more detailed discussion of the importance of using ascending time code.

THE MIDNIGHT PROBLEM

When using the 24-hour clock as a source of time code, a problem arises at midnight, when the clock time changes to 00:00:00:00. If you are attempting to make an edit just past this zero point on the clock, the fifteen-second leader, or preroll time, will start before zero, or somewhere near 23:59:50:00. As the system counts down to the edit point from 23:59:50:00, it crosses over 00:00:00:00. This confuses the system, as it thinks it must rewind toward zero. The system will then stop and try to recue again and again. Although this problem does not occur very often, most production personnel are aware of the problem and will usually take a short break to wait until the 24 hour clock resets to zero, then will resume taping.

If a tape has been recorded with this problem, the editor can replace a small section of the time code with a more convenient time code number that the system can handle. The edit can then be made normally. The editor must note this in the edit log so that the tape master is assembled accordingly. Some newer editing software takes this problem into account, but most editing systems in use today are not capable of resolving it.

PING-PONGING AUDIO

Ping-ponging refers to the process of switching audio and time code from one channel to another so that a tape can be properly edited. Although this technique is not used much anymore, it is required in some instances.

The accepted standard on older three-quarter-inch VCRs used for computerized editing calls for channel 1 to record and play back program audio and channel 2 to record and play back time code. In some cases, however, these two channels are reversed in the recording process. If the VCR does not include a separate address track for time code, ping-ponging is required to put the program sound and time code on the correct tracks for editing.

In this procedure, the audio on channel 2 is rerecorded on channel 1 in one pass. Then, in a second pass, time code is recorded on channel 2 in place of the old program sound. Correcting tapes in this manner is time-consuming and expensive. In addition, cross talk (interference between the audio and time code channels) often results and is heard as low-frequency buzz on the audio channel. This type of interference is usually found only on three-quarter-inch videocassettes using channel 2 for time code. Another negative result of ping-ponging is that the program sound loses one generation of quality by being rerecorded onto another channel.

CONCLUSION

A little care in recording and duplicating time code will prevent most edit abort errors and speed up post-production. Paying attention to time code from the beginning is the key to avoiding problems in the editing process.

Off-Line and On-Line Videotape Editing

6

Today's rapidly improving technology and high hourly rates for on-line editing have caused many producers to spend more time in off-line editing sessions. Off-line editing involves using work copies of original material to build an edited work tape and generate a set of time code numbers that can be used in an on-line session to conform the master tapes into a finished product suitable for broadcast, duplication, or presentation. The general rule is unless you have a large editing budget or have only a small amount of work to do, direct on-line editing should be avoided.

OFF-LINE EDITING

Regardless of the type of editing system used, the purpose of off-line editing is twofold. First, the off-line process is used to produce a work tape so that program continuity can be evaluated. Second, it is used to generate an edit decision list (EDL) that can be used to create an edited master. This list might be hand-written, stored on paper punch tape or a floppy disk, or printed out by a computer. (Chapter 8 discusses the structure of an edit decision list in detail.)

Early Off-Line Editing

In the early 1970s, television producers used consumer reel-to-reel half-inch VTRs to view production material and select good takes prior to the on-line editing session. (See Figure 6.1.) (This older half-inch format, known as EIAJ, should not be confused with the consumer Beta or VHS formats or the newer professional Betacam and MII formats in use today. The EIAJ format is no longer manufactured.)

Once the program was recorded on a master two-inch videotape, a work copy was made by transferring the picture and sound to the half-inch helical-scan reel-to-reel VTR. Although these recorders were not equipped to read time code, visual time code numbers indexing every television frame were combined with the picture to produce a window dub.

The production staff then viewed and selected the good takes using the half-inch work tape. Later, the same work tape was used to select specific edit points, and a frame-accurate paper edit list was generated prior to the on-line editing session.

Before long dual control track and time code editing systems (Figure 6.2) were introduced, and editors were able to produce edited work tapes to show program continuity. These systems were frame accurate and were a giant step forward. The production staff could see the edits on tape instead of just viewing selected material in an unedited form. This was the inception of off-line editing.

Figure 6.1
Using a half-inch reel-to-reel VTR to find edit points.

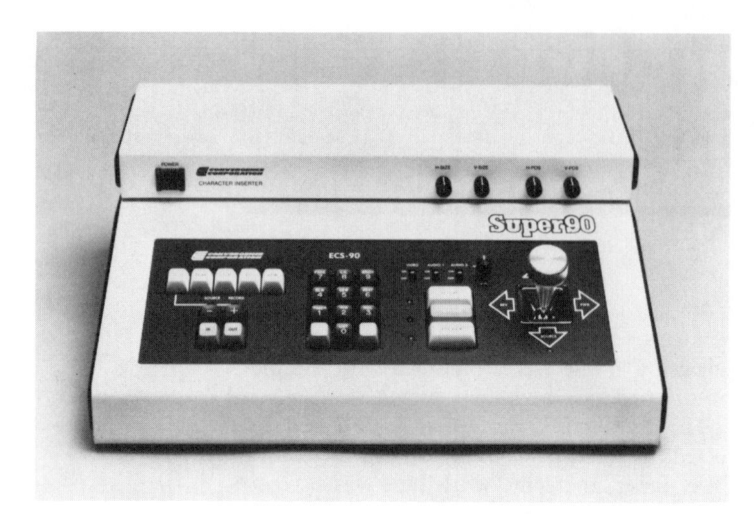

FIGURE 6.2
A low-cost control track or time code videotape editing system with a built-in character inserter for making window dubs. (Courtesy Convergence Corporation)

Computer Assisted Systems

In the mid-1970s, computerized frame-accurate off-line editing systems began to appear. Figure 6.3 shows a typical computer assisted time code–based editing system. In effect these were low-cost versions of their on-line cousins. In addition to being frame accurate, they had the ability to generate optical effects such as dissolves, fades, and wipes. Cueing and searching of time code was a new feature not found on control track editors. Most computerized time code–based editing systems also included a large memory to store and retrieve up to 1,000 edits.

These systems could store edits on a floppy disk or paper punch tape and print out edit lists so that they could be reviewed and modified on paper. Their number-crunching capability helped editors prepare an accurate edit list for an on-line assembly. These systems are the most popular today among videotape editors.

One of the primary differences in cost between computerized off-line and on-line editing systems is the hardware. In many cases, the edit controllers used in both types of system are identical and cost the same, but the VTRs, time base correctors, switchers, and other peripheral hardware required for on-line editing result in a more expensive system.

Most computer assisted editing controllers are dedicated systems designed to perform one function. Calaway Engineering Company of Sierra Madre, California, has taken a different approach to this task. Calaway has designed

FIGURE 6.3
A time code–based off-line editing system. (Courtesy CMX Corporation)

hardware and software that can be used with an IBM PC to edit videotape using low-cost VHS or Beta format VCRs. (See Figure 6.4.) For a small investment, you can turn your IBM PC or PC compatible into a pretty sophisticated editing system. The package comes complete with hardware, software, and a modified keyboard with special color-coded keys displaying editing commands.

FIGURE 6.4
A personal computer used as a videotape editing system. (Courtesy Calaway Engineering)

The Latest Technology

In 1984 two new off-line editing systems, the Montage Picture Processor and the Editdroid, were introduced to the television industry. Previous off-line editing systems were designed to edit in a linear fashion, starting at the beginning of the tape and progressing sequentially through it. These new systems were designed to allow the editor to manipulate edits and juxtapose scenes, sequences, or entire acts without having to rebuild them on another tape. Because they are able to build sequences out of order, they are called nonlinear systems.

Unlike conventional off-line editing systems, the Montage does not create an edited work tape but uses up to 17 Betamax VCRs under computer control to search out and automatically make edits by switching from one machine to another. The Montage is able to store up to 5 hours of unedited material from which you can select segments for the edited master.

The Editdroid uses a random-access editing system based on laser disks, each holding up to 30 minutes of material. Random-access search time from one end of a 30-minute disk to another is less than 3 seconds. Several laser disk players can be added to extend the Editdroid's capacity.

In 1985 a third system, the Ediflex, made its appearance. (See Figure 6.5.) This system uses eight VHS VCRs in a fashion similar to the way the Montage uses Betamax VCRs. To operate the Ediflex, the editor uses a light pen to access various editing functions displayed on the monitor. The Ediflex can hold up to 1 hour of unedited production material.

A year later the CMX Corporation unveiled its CMX-6000 laser disk–based editing system (Figure 6.6). A relative of the earlier CMX-600 editing system, the 6000 provides fast, accurate video editing at a fraction of the cost of the older 600. Although a relative newcomer to the editing community, it has already become a popular post-production tool among both film and video editors.

The Montage, Editdroid, and CMX-6000 are available for sale or lease; the Ediflex can be rented or leased but is not for sale. All four systems have one thing in common: They are able to make edits in a nonlinear fashion. Scenes can be edited out of sequence and then rearranged in any manner the editor desires. Since no work tape is created during the editing process, changes are quite easy to make. A videotape copy of any portion of the edited material can be transferred to a VCR and given to the production staff for viewing.

These systems also are designed to edit single camera productions shot on film and transferred to tape for editing, as well as programs originated on videotape. They include the option of converting time code into film edge numbers (also called key numbers), allowing the producer to conform the original camera film and release his or her production on the motion picture screen. This results in reduced post-production time, and the sooner the film is released, the sooner income starts flowing in.

FIGURE 6.5
The Ediflex random-access videotape editing system for episodic or dramatic series. (Courtesy Cinedco)

FIGURE 6.6
The CMX-6000 laser disk–based videotape editing system. (Courtesy CMX Corporation)

Rent, Lease, or Buy?

The producer or director has three options when choosing an off-line editing system. First, he or she can rent a system. Rental rates in the late 1980s ranged from about $20 an hour for a very simple cuts-only, two-VTR editing system without time code and without an editor (add $15 to $50 an hour for an editor) to about $175 for a full-blown system with time code, four or five VTRs, an effects switcher, a time base corrector, and an editor.

FIGURE 6.7
An example of an on-line one-inch editing suite.

The second option is leasing. Leasing might be more cost-effective than renting in the long run. In addition, the producer or director has the use of the system 24 hours a day, 7 days a week. Leasing on a long-term basis can reduce off-line editing costs significantly.

Finally, a producer or director can buy a system. One advantage of buying a system is that he or she can preview material at home or in the office without having to travel to an editing facility at a specified time. A complete low-cost off-line cuts only editing system with two VCRs, monitors, cables, and all the necessary hardware can be had for around $12,000, while a highly sophisticated system might cost as much as $125,000.

In the end, the decision of whether to rent, lease, or buy comes down to how much you plan to use the system, what type of system you need, and how much you are willing to pay for the convenience of having the system at your fingertips.

ON-LINE EDITING

On-line editing generally refers to the assembly process in which original unedited material is conformed to a master edited tape suitable for viewing, presentation, or broadcast. On-line editing also might include the production of complex digital effects and titles not included in the final work tape. Figure 6.7 shows a typical on-line editing suite.

A master edited videotape can be assembled automatically by feeding in an EDL, which the computer uses to locate material. The tape operator only has to change reels in accordance with instructions from the computer program. This process is analogous to film negative cutting, which is also a number-matching process used to conform original film material to an edited film work print. The edited master tape also can be assembled manually or by using a combination of both methods. Of course, a fully automatic assembly is the most efficient and cost-effective method of assembly.

Since the on-line editing process generally involves expensive and very complex equipment, it is prudent to be as well prepared as possible. Editing charges generally continue to add up even if the crew in the editing room has to wait while the editor or production staff solves any problems.

When scheduling your on-line session, make sure the facility can accept your EDL from a floppy disk or paper punch tape. Although paper tape edit lists used to be universally accepted, many facilities now use floppy disks instead. Don't assume that your list will be compatible with a particular facility's equipment. Many facilities can now convert 3½-inch or 5¼-inch floppy disks generated on a personal computer to 8 inch industry standard formats, or even use the small disks directly without conversion.

The On-Line Editor

An on-line editor must be creative, have an in-depth understanding of audio and video signals, and be familiar with the operation of sophisticated switchers, digital effects devices, and specialized hardware and software. The editor also must make sure that all the equipment he or she will need during the editing session is ready to go prior to the start of the session. This prevents confusion and expensive on-line charges for time spent making the equipment operational.

Videotape Standards

The master videotape must meet certain FCC technical requirements if it will be used for television. These standards do not apply to image quality but only to the technical parameters of the video signal itself. Some of these include audio and video levels and horizontal and vertical blanking widths (relating to image size on the screen).

Material from sources such as home VCRs; quarter-inch, half-inch, or three-quarter-inch cassette copies; and other low-resolution consumer video and audio formats is generally unsuitable for professional use. If this material is to be used on television, it often must be processed by electronic equipment such as time base correctors (TBCs), noise reducers, frame stores, and image enhancers before it can be recorded on another videotape or broadcast. Most of the hardware needed to prepare these small format tapes for broadcast is quite expensive. For example, a TBC might cost anywhere from a few thousand dollars to more than $25,000. The Sony BVX-30 combines a TBC, an image enhancer, a noise reducer, and a color corrector in one unit and costs close to $50,000.

Many of the sporting events, news shows, and documentaries we view on television are recorded on small format VCRs using the three-quarter-inch U-matic or half-inch Betacam or MII equipment and tape. First-run sitcoms, movies of the week, and feature films all use the higher-quality one-inch format. Although the term *on-line* generally refers to the editing of a one-inch master, today virtually any format can be used to generate a master tape as long as it meets FCC broadcast specifications.

On-Line Tasks and Rates

Because of the high cost of on-line editing, it is not advisable to make editing changes during on-line sessions. The only time changes are made is if an obvious editing problem was missed during the off-line process.

Titles are generally added during the on-line session. In addition, complex optical effects are inserted into the program. Often these are built in a separate session and merely inserted into the tape during the assembly process.

On-line editing rates run between $275 per hour for the simplest two-VTR system to $1,500 or more for a system that includes five one-inch VTRs and the equipment needed to create complex digital optical effects and graphics and to perform audio mixing.

The main reason video on-line editing is so expensive is the capital investment in equipment. For instance, a fully equipped one-inch tape machine costs more than $100,000, a sophisticated video switcher costs about $300,000, and a digital effects generator costs as much as $350,000 per channel. With this equipment, however, the possibilities for videotape editing are virtually endless.

OFF-LINE VERSUS ON-LINE EDITING

Advantages of Off-Line Editing

1. It costs less than on-line editing.
2. It is faster than film editing, since it is not labor intensive.
3. Because the cost is low, you can edit several different versions.
4. There is less pressure on the client and the editor to get it right the first time.
5. You can create many standard optical effects.
6. You can store edits for the on-line session.
7. The cost of editing stock is low.
8. You can make an approval copy of the edited tape to send to the client.

Disadvantages of Off-Line Editing

1. Three-quarter-inch VCRs move tape more slowly than one-inch VTRs.
2. Excessive time spent in an off-line session or "polishing" your program might increase the editing budget.
3. Special equipment to create digital effects is generally not available.
4. The work tape does not always reflect the quality of the master tape.

Advantages of On-Line Editing

1. Most new one-inch VTRs wind at 50 times play speed, increasing the overall efficiency of an automatic assembly.
2. Many digital optical effects are available.
3. Color correction can be used to touch up material.
4. Multifaceted graphics and titles can be created.
5. The final product can be touched up as it is assembled.
6. The end product is a high-quality edited master suitable for broadcast, duplication, or distribution.

Disadvantages of On-Line Editing

1. Hourly rates are high.
2. Additional electronic devices might not be included in the base price.
3. You must be prepared to begin editing immediately, or the cost will reflect your lack of organization.

CONCLUSION

Computer assisted videotape editing is faster than manual editing, repeatable, and frame accurate. Because of these advantages, it provides much more flexibility than other forms of editing.

The off-line editing system EDL you create must be compatible with the on-line system. Only then will you be able to take advantage of relatively low cost off-line systems to produce a work tape that can be used in the on-line assembly process. Remember that off-line mistakes must be corrected at on-line rates, which increases the cost of the editing process. Thus it is important to correct as many mistakes as possible in the off-line session.

Today's technology is more advanced and complicated than ever before. New editing tools allow the editor to minimize the amount of bookkeeping he or she must do, to be more efficient, and, in turn, be more creative throughout the editing process.

Tape Preparation for Editing 7

A new roll of videotape is unsuitable for editing because it is blank. For you to perform any kind of electronic editing, the tape must contain the proper video and synchronizing signals. The term *edit black* refers to a tape that has been prepared for editing in this way. Other terms used to describe such a tape are base black, crystal black, prestriped, color bar encoded, or just coded. For electronic editing systems to make clean, glitch-free edits, a tape must be prerecorded with the necessary signals for its entire length in advance of the editing session.

Cleaning the VTR or VCR before recording a videotape also is an important function. This applies to consumer as well as professional equipment. Accumulated tape oxide on the video and audio heads degrades the quality of the recorded signal. Chapter 9 outlines the proper procedures for cleaning equipment and handling videotape.

BASIC TAPE SIGNALS

Every videotape used for broadcast or in the home must include a control track signal and a solid synchronous reference video signal. A tape also might have the following optional signals: longitudinal time code, vertical interval time code (VITC), or an audio tone.

The control track signal maintains a constant tape speed through the VTR or VCR. Control track pulses are analogous to film sprocket holes.

The reference video signal can be any stable signal, such as program material, color bars, viedo black, or a live television camera. The video scanner of a VTR or VCR must lock onto the signal so that it can display a stable image.

Time code can be recorded on a longitudinal audio channel or within the vertical interval of the video signal. Although it is optional, it is necessary if you expect to maintain frame accuracy and repeatability. It also indexes and gives you a reference point for every frame on the tape. Time code is discussed in detail in Chapter 5.

The VTR or VCR being used must be capable of recording time code on an audio channel or in the vertical interval of the video signal. If the VTR or VCR has only one audio channel, time code cannot be recorded because it will erase all program audio. Many low-cost home VCRs fall into this category. Generally, only professional VCRs are equipped to record time code on one of two or three separate audio channels with enough fidelity to be decoded by time code reading equipment.

When you copy a videotape from one VCR to another, you are making a base recording in the same way you would during the transfer editing process (discussed later in this chapter), only sound and pictures are being recorded instead of the base black or color bars. During the duplicating process, all the required synchronizing signals are recorded along with the sound and picture information. Although no editing takes place, the basic time code and synchronizing signals being recorded on the blank videotape are identical to those on the source tape.

RECORDING EDIT BLACK

When a VCR or VTR is put into record, the video signal and the control track pulses are being recorded on the raw tape. The time code signal can be recorded at the same time.

Most professional editors start recording the base signal at the head of the videotape, presetting the time code generator to 00:57:00:00. Three minutes later, the time code will change to 01:00:00:00, which can be used as a zero reference point for timing the length of the program. These three minutes are then used to record at least one and a half minutes of reference color bars, audio tones to check the level and frequency response of the recording, and an identifying slate. The remainder of this time is recorded as video black prior to the start of the program. Any time code number that allows the program to start at a zero reference may be used. For example, you could start at 00:58:00:00, leaving only two minutes of space before 01:00:00:00, or at 09:58:00:00, which would read 10:00:00:00 after two minutes.

Most raw videotape stock contains several minutes more than the designated length. For instance, a 30-minute tape generally contains about 34 minutes, allowing for the extra 3 minutes of test signals and an extra minute at the end of the program. A 60-minute tape might contain up to 66 minutes, and a 90-minute tape might contain 96 minutes. Even short reels or cassettes of about 10 minutes contain at least 2 extra minutes. The amount of extra tape included varies from one manufacturer to another.

When recording time code in the NTSC television standard, it is very important to set the time code generator to drop frame or non–drop frame. When using a time code–based computer assisted editing system, all computations are based on the type of time code recorded on the edit black tape. Mixing the two types of time code is not recommended because computation errors and confusion might result.

During the editing process, the black or color bar signal or the old program material that comprises the base recording is erased, and the new program material replaces it on a frame-for-frame basis. Although it is not mandatory to record an audio tone as part of the base signal, doing so helps set record and playback levels. An audio tone should be recorded the length of the color bars for convenience.

Post-production facilities generally use a header format on their edit black videotape, which allows for test signals, a slate, and sometimes space for promos. Although there is no specific industry standard, most facilities try to adhere to the following format:

Matrix or other special test signals
Color bars for at least one minute with
  white noise or tone as a reference for sound
Identification of left or right channel if stereo
Video black as a spacer for slates and promos
Program content
At least one minute of black following the program

INSERT EDITING

Insert editing is a process used to build an edited master tape on videotape stock that has been prerecorded with the proper signals. In other words, insert editing means selectively copying video and/or audio to precoded videotape without interrupting the control track signal or time code. Today virtually all electronic editing is done using insert editing.

In other words, the electronic process used to copy signals from one VTR or VCR to another during editing is known as transfer editing—that is, the video and audio signals are transferred from one machine acting as a playback source through electronic circuits to another machine containing a preblacked videotape, which is recording those signals. If care is taken in preparing the base tape, the resulting edited master should play properly on any compatible VCR or VTR.

Another type of time code may be recorded as part of the video signal and is called VITC (vertical interval time code). When time code is recorded in the video signal, it is actually recorded in the vertical interval portion of the signal and not as part of the visible picture on the screen. If time code is recorded in this manner, it should also be recorded longitudinally on another track since the VITC information cannot be read in a high-speed search. Longitudinal time code is preferred when videotapes are played at either fast-forward or rewind speeds, while VITC is quite accurate when editing at or below play speed and is frame accurate even when the tape is in still-frame mode.

VITC is very valuable in locating edit points precisely. When the VTR is stopped and the desired frame is displayed, the editor just marks the edit point, entering the eight-digit time code number into the system's memory. The only catch is that your VCR and editing system must be capable of coding and decoding this signal. In addition, VITC will not work with any VCR unless the VCR is equipped with dynamic tracking hardware, which stabilizes the video so that the video scanner is able to read VITC reliably.

VITC is not standard equipment on all VTRs or VCRs. Sony uses the term *dynamic tracking* (DT) to describe its VITC hardware, while Ampex calls this hardware *auto scan tracking* (AST).

VITC is helpful on tapes used as source material or on the edited master tape to locate edit points even when the tape is motionless, but the source VITC time code is not duplicated when material is copied to the edited master. When an edit is made by transferring material from the source VTR to the master record VTR, the resulting VITC time code on the edited master reflects the same time code information in the VITC area as on the corresponding longitudinal time code track on the edit master tape. In other words, the videotape master recorder inserts the VITC time code into the video signal on the edited master tape, which came from the longitudinal time code information on the edited master tape.

VITC time code is applied to a videotape at the time the material is recorded on the tape and cannot be erased or re-recorded since it becomes part of the video synchronizing signal, which is not copied from generation to generation during editing. Generally this occurs during studio production or a telecine transfer. VITC is usually not recorded in the field unless it will be required for editing later on. The main reason for this is the extra cost associated with the hardware. Lines 12 and 14 of the vertical interval area of the video signal are generally used to record this information. See Chapter 5 for more information about VITC.

VITC cannot be added to the video information once it has been recorded on the tape. You can, however, add longitudinal time code to the tape after material has been recorded, although doing so is more time-consuming and costly than recording it at the same time the production material is being recorded. Adding time code to videotapes recorded in the field is acceptable if a time code generator is not available in the field. In this case, time code is added later to the time code track, which is channel 2 (on older VCRs) or channel 3 (on modern one-inch VTRs).

Time code cannot be added to the address track on VCRs once the video has been recorded because it erases part of the video signal. This is not true for one-inch VTRs, which use a different recording process that allows you to add time code to the address track at any time.

You should not try to add longitudinal time code after the program picture and sound have been recorded if the production is designed as a multiple camera shoot. In that case, all the VTRs must have the same time code information recorded on all the tapes at the same time during production. Trying to add time code to each tape later and then synchronize them accurately is a very difficult, costly, and time-consuming procedure.

ASSEMBLE EDITING

Assemble editing does not require a prerecorded base for the entire length of the tape. Only a short amount (30 to 60 seconds) of black, color bars, or other synchronous video signal is recorded manually at the head of the tape. This generates the same kind of signals on the tape as those recorded for base black.

In assemble editing, this first short piece of video and control track is used to stabilize the system so that when the VTR or VCR begins recording at the edit point, new control track and video, along with audio and time code, are edited to the signals previously recorded on the tape. This add-on procedure occurs every time the VCR or VTR goes into record mode.

Assemble editing can be used to make noncritical edits for a presentation when no edit black is available. If raw videotape stock is used in the assemble mode, however, defects on the tape might not be discovered until it is too late. Electronic evaluators are one way to scan the tape for obvious defects. Another method is to view the tape in real time from beginning to end, which is the most effective way to determine overall quality, but it is time consuming and costly since it ties up a machine and an operator to view a tape through its entire length in real time.

Edit List Management

8

Computer assisted videotape editing in any form deals exclusively with time code and frame numbers. Managing these numbers is a problem editors face every day. Edit decision lists (EDLs) are the principal means of documenting and storing the time code numbers generated during the videotape editing process.

CREATING AND RECORDING AN EDIT

The method used most often in videotape editing is known as transfer editing. This simply means that material from a source tape is transferred to (recorded on) another tape by an electronic copying process.

The beginning and end points of an edit are identified by time code, which consists of eight numbers grouped in a specific way. These numbers represent the hour, minute, second, and frame. The following example of an eight-digit time code number represents a point at 5 hours, 17 minutes, 16 seconds, and 0 frames: 05:17:16:00.

All time code generators are based on a 24-hour clock, and the highest time code number possible is 23:59:59:29—23 hours, 59 minutes, 59 seconds, and 29 frames. After that, the generator automatically starts counting over at zero.

The colons in the time code number are called separators or delimiters. When you are making an edit, the first time code number is stored in the computer's memory as the beginning of the edit. The second time code number, which marks the end of the edit, must always be larger than the first. The difference between them is known as the edit duration. For example, an edit beginning at 05:17:16:00 and ending at 05:17:26:01 would have a duration of 10:01 (10 seconds and 1 frame). This is how the edit is stored in the EDL.

Assuming that the edit just described represents the time code numbers of the source material, you must now determine where that edit will be placed on the master edited tape. The source tape and master edited tape each has its own set of time code numbers. To record the edit from the source tape onto the master tape, you must mark the edit.

To mark an edit on the source VTR, you just hit one key (Mark In) on the edit controller keyboard, which enters all eight digits of the beginning time code number simultaneously. You can mark the end of the edit in a similar manner using the Mark Out key. To determine the placement of that edit on the record tape, you need only mark the starting point, and the computer will automatically calculate the end point based on the duration of the edit on the source tape.

Conversely, you might want to mark the beginning and end points on the master record tape and fill that hole with material from a source VTR. The procedure is the same as when you mark the in and out points on the source VTR. If you mark an in and out edit point on the record VTR and then locate the beginning of the source material, the computer will automatically calculate the source out time based on the edit duration on the master tape. Here is an example of Source-In and Out times and Record-In and Out times:

Source-In	Source-Out
05:17:16:00	05:17:26:01

Duration = 10:01

Record-In	Record-Out
01:00:00:00	01:00:10:01

Duration: 10:01

Notice that the edit durations are the same. This enables the editor to modify or correct edits easily.

In addition to the source and master time code, other pertinent information is stored in the computer. This includes the edit (or event) number, the reel number, the edit mode (audio and video, video only, or audio only), and the transition type (cut, dissolve, wipe, or other optical effect). These items are discussed in more detail later in this chapter. All this edit information, along with any appropriate comments, is stored event by event in the computer's memory or on one of several types of storage devices, including paper punch tape or a floppy disk.

LINEAR VERSUS NONLINEAR EDITING

Two methods of editing are used to generate an EDL. The first and most common type is linear editing, which is the building of edits sequentially down the length of the tape. Linear editing generates the type of list I discuss throughout this chapter.

The second method of editing is called nonlinear editing. In nonlinear editing, the editor selects the exact picture and/or sound information for each edit as it is displayed in real time on several (up to 17) half-inch VCRs. The editor marks the beginning and end points of each edit on the computer, then views the edits as they are played back according to the information stored in the computer. The important point to remember is that the edits are not recorded on videotape. Instead, the edit information is stored in the computer's memory and can be modified quickly since no rerecording of edits is required.

One advantage of nonlinear editing systems is that the edit list created by the computer is frame accurate and by design contains only correct edits in the proper sequence. No list management is required because the edits, even after being modified many times, are stored in the correct form. This is important since once a series of edits has been generated in the play or viewing mode, the system will play the edits correctly only if the edit list stored in memory, which was used to generate the edits, is correct. In other words, unless the edit list in memory represents what the editor built, the results will not reflect what is displayed on the edit monitor.

Although nonlinear editing systems are relatively new in post-production, it is apparent that they will have a substantial impact on the editing of television programs and feature films for years to come. Some of these nonlinear editing systems are discussed in Chapter 14.

STORING TIME CODE DATA

Although time code editing was introduced in 1967, it was not until early 1972 that editors began to concern themselves with edit list problems. The reason for this was that prior to 1972, no way of storing time code information in memory was available.

Once information storage became available, editors realized that a means of retrieval was necessary to repeat or locate a given edit on videotape. Interfacing a teletype reader and paper tape perforator (punch) to a computer editing system was a simple and inexpensive way of storing editing data for future use. The paper punch tape could be read into a teletype terminal at a later date, yielding an edit list that contained all the edit information stored on the tape.

Even though storage on paper tape is still quite popular today, the large number of edits required by videotape limits its usefulness. Only about 250 edits can be stored on paper tape before the tape becomes cumbersome and the reader or punch mechanism jams.

In the early 1980s, paper tape storage of editing data was gradually replaced by floppy disks. One disk, which is about the size of a 45 rpm record, can store more than 3,000 edits on each side. In addition, disks are relatively inexpensive and can be used again and again.

DEALING WITH NUMBERS

Some editors are able to manipulate frame identification numbers with great ease and with comparatively few edit list errors. Generally, though, flying fingers do not an editor make. There are three major reasons why edit list errors occur:

1. The editor is inexperienced or not familiar with the system.
2. The editor does not understand the relationship between the code numbers and the edit list.
3. The editor keyboards the data too fast and does not verify the data as it is entered.

It takes only a few seconds to check each entry during the keyboarding operation, especially when working from a previously created list. Repairing or correcting erroneous numbers in an already completed edit list takes longer and becomes more difficult. The editor must always remember that the computer and its memory will store only the information that the editor feeds it, and that includes the editor's errors.

A fourth source of errors in computerized editing systems is a power surge or transient. These surges can be suppressed by installing a voltage protection device. The cost of such a device is low, and using it can save you a lot of headaches in the long run.

DESCRIBING AN EDIT

Figure 8.1 is an example of a typical entry in an edit list. The entry "001" in column *a* is the sequential edit (or event) number. It increases by one for each consecutive edit entered. At the end of the list, it gives the total number of edits in the list. This number is continuously updated as edits are added to the edit list.

The entry "63" in column *b* is the original reel number associated with this edit. Each edit must be assigned an original reel number to allow the system to locate the proper source material. Some older editing systems can handle only 63 reel numbers, while many newer systems can handle six-digit reel numbers or any six-character alphanumeric combination.

The entry "B D" in column *c* identifies the edit mode and the type of transition. Three edit modes are generally possible. The letter B indicates a both cut, meaning that both the video and the audio are cut and changed simultaneously. The letter V indicates that the edit applies only to the video (the picture), leaving the audio on the destination tape intact. The letter A indicates that the edit applies only to the audio (the sound), leaving the video intact.

One of four letters is used to indicate the type of transition desired. The letter C indicates a straight, or direct, cut that will produce an abrupt transition from one scene to the next. The letter D indicates a dissolve (superimposed fade-in and fade-out). The letter W indicates a wipe (a transition that crosses the screen in any direction with a visible hard, soft, or bordered edge). Finally, the letter K indicates the use of a key. A key is an electronic method of combining, or matting, one signal on top of another without allowing any of the background to bleed through.

a	b	c	d	e	f	g	h
				Play-In	Play-Out	Rec-in	Rec-Out
001	63	B D	030	12:11:45:00	12:11:55:00	01:00:00:00	01:00:10:00

FIGURE 8.1
A typical edit entry in an edit list.

The three digits in column *d*, "030," indicate the duration of the transition in frames. In Figure 8.1, a both dissolve with a duration of 30 frames is indicated. The maximum duration that can be entered with this system is 999 frames (about 33 seconds at 30 frames per second). A "000" entry or no entry at all in column *d* indicates a direct cut of type C. Transitions of type D, W, or K must always be accompanied by a corresponding duration entry in column *d*.

Columns *e* and *f* indicate the extension of the useful action. The useful action is that part of an original scene (or take) selected by the director or the editor for final program use. The head entry in column *e*, "Play-In," indicates the start of the useful action, and the numbers underneath it indicate exactly where the first frame of the useful action is located on the original source reel. The time code location of this frame is 12 hours, 11 minutes, 45 seconds, and 0 frames. In a similar fashion, the Play-Out heading in column *f* indicates the end of the useful action (12 hours, 11 minutes, 55 seconds, and 0 frames). Note that in this example, the useful action has a length of exactly 10 seconds and 0 frames.

Columns *g* and *h* ("Record-In" and "Record-Out," respectively) define where on the edited tape the original material has been recorded. Here, Record-In defines the first frame of the recorded action and Record-Out defines the last frame of the recorded action.

AVOIDING EDIT LIST ERRORS

The terms *conforming* and *auto-assembly* refer to the important operation of matching the information resulting from a specific editing session to identical information usually found on a master tape. To put it another way, conforming means matching the sequence of edits made in a work print to the diverse original source materials on a frame-for-frame basis.

Much of today's videotape editing is done off-line. In off-line editing, a work print is created instead of editing directly on the final master tape. The off-line editor should try to minimize editing errors for a number of reasons:

1. Errors might make the off-line or on-line editor feel frustrated with his or her tools and undermine the confidence needed to edit efficiently.
2. The producer or director might feel that the editing system does not provide the necessary flexibility to generate the desired results.
3. If the editor does not deliver a clean edit list prior to conforming or auto-assembly, needless delays result. These delays are caused in part by an excessive number of reel changes and by redundant recordings that waste time and money.

A dirty edit list also can result in on-line editing errors and cause confusion. Additional charges incurred because of edit list errors are generally absorbed by the post-production facility since few if any clients are willing to pay for editing time to correct such errors.

Some systems have the ability to clean the edit list during the off-line or on-line editing process. The editor also can take advantage of a variety of commercially available list-cleaning programs. These programs generally provide the following features:

1. They remove overrecordings (a term explained later in this chapter) so that the audio and video edits are joined in a consecutive manner.
2. They rearrange edits according to Record-In times so that the automatic assembly can proceed in an orderly fashion. This is important because out-of-sequence editing is a common practice among videotape editors. Additional edits, called inserts, also are often added to the edit list after the basic continuity of the program has been established. Editors rely on this software to rearrange the edits rather than performing this task manually and risking the possibility of generating some error.
3. They renumber all edits consecutively and generate a second set of numbers that correspond to the original sequence of edits.
4. They position the separate video and audio edits in proper relation in the edit list based on their Record-In times, no matter where in the list the edit was actually inserted.
5. They trace reel number and edit information through several versions of the edited work print and correctly relate it to the final edited material.

AUTOMATIC LIST MANAGEMENT

Although there are several software tools designed to assist the videotape editor in edit list management, two outstanding programs are 409 and Trace.* These two programs have become the de facto standards for the television industry both in the United States and abroad. The basic principles of these programs are the same as those used in other products of this type. The primary differences are in the ways data is manipulated internally. One of the reasons this type of software is used is to reduce auto-assembly time, which is charged at an on-line rate ranging from about $300 to more than $1,500 an hour.

409 and Trace were introduced to the post-production industry in the late 1970s and are designed for editors who are generally familiar with time code edit lists. Software tools of this type can, however, be used with simple cuts-only or fully configured off-line or on-line editing systems. The only requirement is that the system be capable of generating an edit list and storing it on paper punch tape or a floppy disk in an industry-compatible form. The primary difference among edit list formats is the manner in which the audio channels are designated. For example, one system might use A and AA to describe audio channels 1 and 2, while another might use A1 and A2. Although there are many edit list formats, CMX is one commonly used industry standard. Many standards provide a conversion option that will generate a CMX-compatible list.

List Cleaning

List-cleaning programs such as 409 are designed to be used with time code edit lists stored in computer assisted editing systems. They clean a list by removing overrecorded time codes, deleting certain types of duplicate edits, and resequencing edits in a manner that will allow an efficient automatic assembly. 409 and other programs also provide a number of special options designed for more specific purposes, allowing greater flexibility in cleaning lists.

*Software computer programs are products of the Grass Valley Group.

To use a list-cleaning program, follow these basic steps:

1. Load and run the program.
2. Answer the questions as they are asked.
3. Select the input and output formats if required.
4. Instruct the computer to read the file number from the floppy disk or paper tape.
5. Select either a checkerboard cleanup (B mode) or sequential assembly cleanup (A mode).

If your system is equipped with a printer, you should be able to print out the cleaned list. 409 allows you to print out your list before cleaning, providing you with a copy of the list as it was entered into the computer. If the program deletes an edit you did not want to delete, you can check the original list and correct the problem. (These programs seldom if ever delete data unless it was entered incorrectly, but the printout allows you to catch your own entry errors as well.)

Trace Programs

Trace programs such as Trace backtrack through a sequence of edit lists and assign priorities based on the most recent list. The program then generates a final edit list incorporating all the editing generations. Trace enables you to make a series of changes without having to rebuild a single master work tape each time a new generation of changes is made. This saves time, effort, and money.

Following is an example of a typical Trace procedure. The first edited cut (cut-1) of a program is mounted in a play VCR and is used as a new input source by assigning it a reel number not already used in a previous list. This first cut is copied onto another coded work tape. Any part of the source tape can then be copied onto the new coded work tape in any order. New material can be added, or existing material can be rearranged, thus creating another edited work tape. This is called the second cut, or cut-2, and it generates its own edit list, which includes sections of the first cut plus any new or copied material that might have been added.

The process of using each edited work tape as a source tape to create a new version can continue in this fashion up to ten generations. The only consideration is that the edited work tapes used in the tracing process be assigned new reel numbers that are not in conflict with those already in use in the original edit list.

The last edit list revision is always called the final cut, which corresponds to the final version of the edited work tape. All the versions are entered into Trace, with the final version entered last. In a few minutes, depending on the number of edits stored, the list is completely traced, resulting in a single edit list reflecting the continuity of the final work tape. This traced edit list can then be used to conform another master tape reflecting only those audio and video edits that were used to create the final version of the work tape.

POTENTIAL PROBLEMS

Certain types of edits can cause problems with your edit list. These problems and possible solutions are explained in the remainder of this chapter.

Overrecordings

A frequent error is overrecording material. During the first assembly of a program, editors generally overrecord material and will later determine where it will cut to the next scene. This extra material at the end of a scene remains in the edit list and is eventually covered up by new material when the cut is added. Leaving the extra material in the edit list reduces the efficiency of any auto-assembly.

Figure 8.2 is a typical example of an overrecording left in an edit list. Note that the Record-Out time of edit 001 (last column) is five seconds longer than the Record-In time of edit 002 (next to last column). This is a short overrecord, but overrecords can amount to several minutes of extra material for each edit. In auto-assembly, the extra material is recorded, the reel is rewound, and the new scene is recorded over the material. If the Record-Out time of edit 001 is shortened to match the Record-In time of edit 002, this step is eliminated in auto-assembly. Figure 8.3 shows what the cleaned edit list would look like.

Another frequent overrecording error arises when two consecutive both cuts are straddled

FIGURE 8.2
An example of an overrecording left in an edit list.

001	63	B	C	12:11:45:00	12:11:55:00	01:00:00:00	01:00:10:00
002	45	B	C	09:44:15:00	09:44:30:00	01:00:05:00	01:00:20:00

001	63	B	C	12:11:45:00	12:11:50:00	01:00:00:00	01:00:05:00
002	45	B	C	09:44:15:00	09:44:30:00	01:00:05:00	01:00:20:00

FIGURE 8.3
An example of correctly timed consecutive edits in a list.

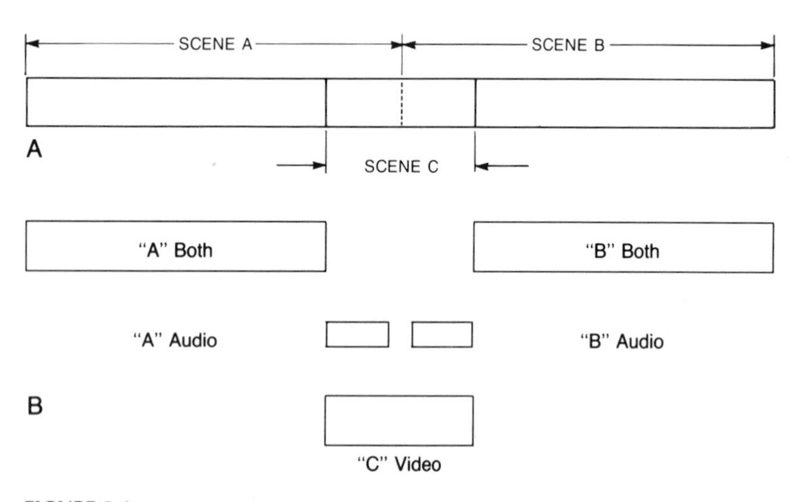

FIGURE 8.4
A schematic of a video insert straddling two both cuts. (A) Uncleaned. (B) Cleaned (by the 409 software program).

001	023	B	C	12:00:15:00	12:00:30:00	01:00:00:00	01:00:15:00
002	017	B	C	08:25:10:00	08:25:20:00	01:00:15:00	01:00:25:00
003	009	V	C	10:00:40:00	10:00:43:00	01:00:14:00	01:00:17:00

FIGURE 8.5
An example of an insert straddling two both cuts.

by a video insert that was added after the initial editing (see Figure 8.4A). The figure reflects the following sequence of transfer recordings: Scene A is recorded first, followed by scene B and then scene C. Figure 8.5 shows the relation of the video insert (scene C) to the both cuts (scenes A and B) in the corresponding edit list.

Assembly of edits 001, 002, and 003 (representing scenes A, B, and C, respectively) must be done exactly in the order indicated in the list. If edit 003 were assembled before edit 002 (which is possible because edit 003's Record-In time is earlier than edit 002's), an error would result. Edit 002 would then be assembled in the last place and would cut off the second half of edit 003. It is not worth the editor's time to rearrange these edits during an off-line session, but a common method of list cleaning would be to break this group of edits into separate audio and video edits, allowing the editing system to record the video scenes in sequence and the audio tracks in sequence in a piecemeal fashion determined by the type of editing system used. This prevents erasing previously recorded material.

Figures 8.4A and 8.4B illustrate two scenes straddled by video insert. A 409 type program would section it as shown. Note that in 8.4A, the edit is made up of three pieces, two both cuts straddled by a video insert. In 8.4B, five edits were generated. Two both cuts, the balance of the audio from scenes A and B and the video insert allowing this cleaned version to be assembled out of sequence without erasing material as might occur in 8.4A if the three edits were assembled by record in times, i.e., A, C, B.

Tracking Edits

A tracking edit is a nonrecorded, redundant edit. As shown in Figure 8.6, its Play-In and Play-Out times are identical, and thus it has a zero duration. The only purpose of a tracking edit is to provide synchronizing information for the computer, enabling it to keep the audio or video in sync with the master while it generates a special effect. A tracking edit must precede an edit that creates an optical effect such as a wipe, dissolve, fade, or key. The tracking edit contains the Play-Out and Record-Out times of the specific edit from which the editor wishes to start a given effect.

A tracking edit must be included in the edit list when the record times of consecutive effects are separated by one or more audio or video edits not relating directly to the previous event. A tracking edit is not required between straight cuts. It is needed only for bookkeeping purposes and is not recorded by the system because it has no duration.

Figure 8.7 illustrates another use of a tracking edit. Notice that there are two consecutive edits identified as edit 004. The first edit is the tracking edit and the second is the destination edit. The editing software recognizes this combination and thus assembles the edits accordingly. If the tracking edit were not included, the software would not assemble the edit properly.

The group of edits in Figure 8.7 allows the editor to add narration (contained in edit 004) and two video inserts (edits 002 and 003) to the master scene (edit 001). Let's analyze this list.

004	003	A	C		05:01:13:24	05:01:13:24	01:00:44:01	01:00:44:01

FIGURE 8.6
An example of a tracking edit.

001	003	B	C		05:01:03:24	05:01:33:24	01:00:34:01	01:01:04:01
002	025	V	C		17:08:00:29	17:08:04:28	01:00:40:15	01:00:44:14
003	030	V	C		14:10:12:00	14:10:15:10	01:00:54:14	01:00:57:24
004	003	A	C		05:01:13:24	05:01:13:24	01:00:44:01	01:00:44:01
004	099	A	D	000	10:11:22:17	10:11:42:25	01:00:44:01	01:01:32:27

FIGURE 8.7
An example of how a tracking edit is used.

001	015	B	C	11:00:10:00	11:00:20:00	01:00:00:00	01:00:10:00
002	008	A	C	07:00:01:00	07:00:03:00	01:00:08:00	01:00:10:00
003	008	B	C	07:00:03:00	07:00:15:00	01:00:10:00	01:00:22:00

A

001	015	AA/V	C	11:00:10:00	11:00:20:00	01:00:00:00	01:00:10:00
002	008	AA	C	07:00:01:00	07:00:15:00	01:00:08:00	01:00:22:00
002	008	V	C	07:00:03:00	07:00:15:00	01:00:10:00	01:00:22:00

B

FIGURE 8.8
An example of a split edit delaying video. (A) Grass Valley mode type display. (B) CMX mode type display.

001	015	B	C	11:00:10:00	11:00:20:00	01:00:00:00	01:00:10:00
002	008	V	C	07:00:01:00	07:00:03:00	01:00:08:00	01:00:10:00
003	008	B	C	07:00:03:00	07:00:15:00	01:00:10:00	01:00:22:00

FIGURE 8.9
An example of a split edit delaying audio.

Edit 001 is a both cut and is the master scene for all that follows. As noted, edits 002 and 003 are video inserts that replace, on a frame-for-frame basis, the corresponding frames of the master scene. Edit 002 starts near the beginning of the master scene, while edit 003 goes in ten seconds after edit 002 stops. The tracking edit, the first half of edit 004, is used to track the audio from edit 001, the master scene, up to frame 01:00:44:01. It synchronizes the audio with the video from edit 001 and mixes it from frame 01:00:44:01 on with the new audio from the second part of edit 004, the narration.

The audio dissolve duration of 000 frames for the second part of edit 004 is merely a means of running the two audio sources together. Because the dissolve duration is zero, no automatic mix of audio will actually take place. The editor will mix the audio sources separately, using an audio mixer. (In most cases, audio mixing is done outside the video editing room. The editor generally prepares the sound tracks for mixing but leaves the actual procedure to another facility.)

Note that the Record-Out time of the second part of edit 004, the narration, comes later than the Record-Out time of edit 001, the master scene. This means that the narration will continue over subsequent edits.

Split Edits

A split edit is a transition between edits that anticipates the picture or sound of an incoming edit. For instance, the sound of an incoming scene might start before we see the picture. This is called an audio leading video edit. Conversely, if the picture starts before the sound, it is called a video leading audio edit. A split edit is also known as an L cut, an offset, or a stagger cut. A split edit can be a very effective transition device.

In the EDL, a split edit appears as a two-line event, as shown in lines 002 and 003 in Figure 8.8. Here edit 001 is a both cut with a duration of ten seconds. Edit 002, which is from a different source reel (008), is an audio cut with a Record-In time two seconds earlier than the Record-Out time of edit 001. The duration of edit 002 is two seconds. Edit 003 is a matching both cut. That is, the Play-Out and Record-Out times of edit 002 are the same as the Play-In and Record-In times of edit 003. Edit 003 also comes from the same reel as edit 002.

What this list indicates is that two seconds before the end of edit 001, an audio edit is made having a duration of two seconds, thereby starting the sound from reel 008 before the picture of reel 015 ends. To continue the sound and switch to the picture of reel 008, a matching both cut is made. Figure 8.8 shows that this edit was made by delaying the video. You could get a different effect by delaying the audio (Figure 8.9). In either case, the result is a transition from one scene to another.

Another method of displaying split edits used by two editing systems manufacturers, Grass Valley and CMX, is shown in Figure 8.8. An explanation of Grass Valley's display is as follows. In edit 001, A12V is used instead of the letter B to define an audio and video edit. In this example, the A12 indicates recording of both channels of audio and the V indicates the video. Edit 002 shows the sound leading the video as A12. In the second half of 002, the V indicates the start of picture. CMX is as follows. Edit 001 is the master scene, a both cut. The first part of edit 002 is the audio leading the video. The second half of edit 002 is the video portion of the split edit. It begins two seconds after the sound begins. Since these two parts of the split edit are connected by the same edit number, the software looks at it as in Figure 8.8, edits 2 and 3.

As the example indicates, the split edit is essentially the same for both edit systems except for the method used to identify and display audio and video edits. CMX describes an audio edit on both channels as AA while Grass Valley describes this same function as A12. Grass Valley provides a translation program called XEDL that converts CMX and several other EDL list formats to make them compatible with the Grass Valley EDL format.

It should be noted that since 1972, the mode type B was used to indicate both audio and video edits, while A and V were used to indicate audio and video edits. Although B, A, and V are still used in other editing systems, AA/V, AA, and V are being used by CMX, and A12V, A12, and V are being used by Grass Valley. This trend will continue for many years, undoubtedly causing confusion and incompatibility among computer assisted editing systems.

List errors can occur if the end of the audio or video insert does not coincide in time with the end of the previous edit, leaving a gap or cutting off a portion of this edit.

Many of the new computer assisted editing systems have the ability to preview and record this type of transition in a single pass. With older systems, however, the editor must create this transition in a two-pass configuration by first previewing and recording the audio or video portion (edit 002 in Figures 8.8 and 8.9), then previewing and recording the both cut in a separate pass. This two-pass process tends to interrupt continuity and makes it more difficult for the editor to see whether the edit has the desired effect.

Optical Effects

Edit list errors frequently occur in the creation of optical effects such as dissolves, fades, wipes, and title keys.

Dissolves. Figure 8.10 illustrates an edit list problem associated with a dissolve. Here the duration of the dissolve is longer than the duration of the edit itself. The edit list shows a dissolve of 90 frames (three seconds), but the edit duration is only 60 frames (two seconds). The dissolve will be cut short by 30 frames (one second), and the edit will not assemble properly.

The problem can be solved by extending edit 002 by at least 30 frames (one second) to a Play-Out time of 12:15:23:00 (also adding one second to the Record-Out time) or shortening the dissolve duration to 60 frames (two seconds) or less. When programming an effect, an editor must always make sure that the duration of the edit is equal to or greater than the duration of the effect.

FIGURE 8.10
Erroneous dissolve instructions in an edit list.

001	296	B	C		07:08:08:00	07:08:18:00	01:00:00:00	01:00:10:00
002	345	B	D	090	12:15:20:00	12:15:22:00	01:00:10:00	01:00:12:00

Fades. The edit list for a fade to black (fade-out), which is really a dissolve to black, is shown in Figure 8.11. Here the length of the dissolve equals the duration of the edit. The letters B and D in columns 3 and 4 stand for "both dissolve," which means that the picture will fade to black (BL in column 2) and the audio will fade to silence.

FIGURE 8.11
Edit list instructions for a fade to black (fade-out).

001	023	B	C		14:59:10:20	15:00:01:20	01:00:00:00	01:00:51:00
002	BL	B	D	090	00:00:00:00	00:00:03:00	01:00:51:00	01:00:54:00

If the editor wanted to add one minute to the end of the fade in a single pass, the edit would appear as shown in Figure 8.12. In that case, the 90 frames of the dissolve duration are added to the one minute of black leader, mak-

FIGURE 8.12
An example of a fade with one minute added at the end.

001	023	B	C		14:59:10:00	15:00:01:00	01:00:00:00	01:00:51:00
002	BL	B	D	090	00:00:00:00	00:01:03:00	01:00:51:00	01:01:54:00

ing the total Play-Out time 00:01:03:00. This minute is, of course, also added to the Record-Out time. It might be useful to remember that although the edit list gives all edit durations in the time code format (hour, minute, second, and frame), all effect durations are displayed in the list in frames only.

001	123	B	C		11:23:11:00	11:23:21:00	01:00:00:00	01:00:10:00
002	456	B	W	012	10:09:00:00	10:09:10:00	01:00:10:00	01:00:20:00

FIGURE 8.13
Edit list instructions for a wipe.

Wipes. Wipes pose slightly different problems. All the rules explained previously apply, but the editor also must indicate which wipe pattern is desired. A typical wipe instruction is displayed in Figure 8.13. Here the edit list looks basically the same as for a dissolve or fade. But the addition of a wipe pattern number (012) is essential for proper conforming through a computer assisted editing system.

007	017	B	C			15:00:43:00	15:00:44:00	01:00:53:00	01:00:54:00
007	017	B	K	B		15:00:44:00	15:00:53:00	01:00:54:00	01:01:03:02
007	021	B	K		030	12:00:23:00	12:00:23:00	01:00:54:00	01:00:54:00

FIGURE 8.14
Edit list instructions for the matting of titles or keys.

Titles. In general, titles, or keys, are letters or other artwork matted on top of a background scene without the background image bleeding through. Again, the general rules for effect duration apply, but special handling is required.

The edit list in Figure 8.14 contains three event lines concerning the key effect. These lines carry the same event number (007). The first line identifies the background source. The second line indicates at which point in the background material the effect will start. It picks up on the frame next to the Play-Out time of the previous edit line, as indicated by the same source reel number, 017, which becomes the background from here on. To achieve this, the second line contains the instruction BKB, meaning "both key background." The third line contains a new reel number, 021. This reel contains the foreground letters, and the system sees this as the both key (BK) foreground material. The third line lists a duration of 30 frames, which tells the system that the artwork will fade up (come up to full intensity) in the 30 frames indicated by the time code. The "both" cue is used here because even though the key affects only the picture, both picture and sound may continue as background information. Note also that line 3 actually describes the tracking edit for the onset of the key.

The complexity of the key signal and the various combinations that can be achieved require a more detailed explanation than I can give here. No doubt the explanation of a key sounds complex and it is, compared to other types of edit functions. Key dialogue as described here is not used as often as it has been, since video switchers with memory options have taken over the task of creating keys so this dialogue is being replaced by simpler methods. Therefore, the description of keys is included only for historical reference. The example, however, shows how easy it is to make edit list errors that will result in problems during auto-assembly. The operating manual of the specific video editing system you are using will provide detailed instructions for integrating keys into the edit list.

SUMMARY

Videotape editors must employ careful edit list management techniques. It is very important that list errors be caught before they are compounded in auto-assembly. It takes only a little extra time to prevent errors at their point of origin.

To minimize errors, the editor should scan the list for possible discrepancies before storing it on paper punch tape or on a floppy disk. He or she should remove all overrecordings and duplicate edits and make sure that record times do not leave holes in the list. All fades, wipes, and dissolves must be preceded by a zero-length tracking edit and must contain the proper codes and play and record times.

The editor should know the system he or she is working with, understand the relationship of the numbers in the list to the edits they will produce, never be in too much of a hurry, and make sure that all edits in the final list appear sequentially by record time. If the editing system is not capable of reordering edits by record time, the list must be sent to a facility that has the software needed to clean the list. Such a service is normally provided at a small cost and will yield a well-organized list that is ready for auto-assembly.

The Care and Handling of Videotape

9

Once an image has been committed to videotape, it represents the efforts of hundreds of actors, technicians, and other support people, as well as large amounts of money. Therefore, it pays to make sure only the highest quality videotape stock is used in the production and post-production processes. It is also important to handle videotape with care to prevent damage.

Even if videotape is manufactured to the highest standards, problems do occur. Although testing during the manufacturing process is extensive, defects occasionally slip by. When this occurs, the manufacturer should be notified as soon as possible. The sooner the manufacturer is notified, the sooner the problem will be corrected.

At least half the problems blamed on videotape are the result of faulty or dirty equipment. Bad operational practices also are a significant factor. This chapter outlines various problems associated with videotape and suggests possible solutions.

QUALITY ASSURANCE

The quality of videotape has improved steadily since its introduction in 1956. Extended wear, chroma (color) noise reduction, and a better signal-to-noise ratio (clearer pictures) are some of the more noticeable improvements.

Contrary to what some people believe, the videotape signal does not degrade with age. If videotape is properly cared for and stored in the right environment, there will be no noticeable degradation of picture or sound quality. In fact, videotape stored properly over long periods of time has been found to maintain its quality and retain color better than motion picture film. This is because film is made of dye images, which have a tendency to fade, whereas videotape images are composed of electronic signals stored on a magnetic medium called iron oxide.

THE MANUFACTURING PROCESS

Videotape is manufactured in long rolls called webs, which are usually at least 26 inches wide. When run through a machine called a slitter, the web is cut into strands of tape that are wound onto reels. The tape is then polished and inspected electronically and visually for any defects. Each web is considered a batch and is identified as such. If a problem is detected in the plant or in the field, immediate steps can be taken to recall or correct the entire batch.

EVALUATING TAPE STOCK

The most accurate way to evaluate a piece of tape is to record about a minute's worth of color bars and audio tone for a spot check. When the tape is played back, obvious defects will show up. To evaluate a roll of tape, color bars and audio tone should be recorded over its entire length. A trained operator should then view the tape in real time, noting any picture or sound disturbances. If the disturbances are minor, the color bars and audio tone should be rerecorded in that area and checked again. Although this is the preferred method of tape evaluation, it is quite tedious and costly and only the larger post-production facilities offer this service.

If it is not practical to check an entire roll, you might want to spot record about a minute at a time starting about 3 minutes into the roll and then every 20 minutes thereafter. The reason you should start recording about 3 minutes into the roll is that the first 3 minutes are usually reserved for test signals.

In the past two decades, attempts have been made to develop a machine to perform the tedious task of tape evaluation, but these electronic devices cannot pick up as many defects as the human eye and therefore cannot be regarded as totally reliable and consistent in their evaluation. Many post-production houses use these evaluation devices because of the high cost of visual inspection. Where total quality assurance is mandatory, however, visual inspection remains the most reliable method.

TAPE-RELATED PROBLEMS

The following problems primarily concern one-inch industry-standard reel-to-reel VTRs. Nearly all the comments regarding the care and handling of videotape, as well as the possible technical problems, generally apply to all other formats, however. Because VCRs use videotape enclosed in a cassette, they are treated somewhat differently.

As an editor, you are probably not well versed in the technical aspects of videotape stock and equipment, but you should have enough technical knowledge to pinpoint the trouble area for the maintenance engineer. In some cases, a small adjustment by the technician is all that is required to correct apparent tape problems.

If care is not exercised during every stage of production and post-production, the tape might become permanently damaged. A mechanical or electronic problem might result in poor recording or playback and cause permanent damage to the tape itself due to stretching or transport path misalignment. Damage due to improper handling can be caused by improper storage, a bent reel, a nicked reel flange, or improper winding.

Manufacturing Defects

One type of defect is improper tape slitting during manufacture. Although rare, faulty slitting might show up as a "wow" in the audio tone during playback. An audio wow is a slow, repetitive variation in sound level or frequency caused by speed changes of the VTR or VCR or by the tape wandering from side to side on the audio playback head. Once a defect of this type is detected, the tape should be rejected, since there is no way to correct the problem.

Occasionally, a batch of videotape will have surface defects that manifest themselves as some form of picture disturbance. These defects include pitting, pinholes, scratches, wrinkles, and oxide shedding. Pitting and pinholes might result from a problem with the way the oxide formula was applied to the Mylar base. For example, if one of the spray bars was slightly clogged, the emulsion would not be evenly applied to the base.

Scratches can occur during the tape-polishing operation. Scratches also might be caused by tiny nicks on any transport surface with which the tape comes into contact. These nicks are often due to a dirty VTR. It takes less than three minutes to clean a reel-to-reel VTR or a VCR. It is worth three minutes of your time to prevent permanent scratches.

Wrinkles may be caused by a slitting problem. Winding a badly slit reel might create tiny wrinkles on the surface of the tape. Oxide shedding might be caused by a bad binding agent (that is, the material that holds the oxide formula on the Mylar base). If you encounter this problem, check the batch number of each problem roll. If all the rolls are from the same batch, notify the manufacturer immediately.

Stiction

A special antistatic backing is applied to videotape to prevent static buildup, or stiction. During manufacture, after the web has been slit, each strand of tape is wound tightly onto separate metal reels. The antistatic backing separates the base side of the tape from direct contact with the emulsion surface of an adjacent layer, thus preventing static buildup.

Without this backing, static can cause the tape to stick to the highly polished surfaces of the tape transport mechanism and prevent it from moving at all. Stiction also can occur as the videotape passes across the video drum, or scanner, and audio stacks. The remedy for tape stuck in the transport mechanism or in these other places is to reduce the tension on both reels, then carefully loosen the tape by hand.

If stiction continues to occur, the tape should be carefully removed from the transport path and all tape contact surfaces should be cleaned with an approved nontoxic chemical cleaner such as Freon TF. The tape should then be rethreaded through the machine. It is a good idea to clean the VTR each time a new roll of tape is mounted to prevent this problem.

The antistatic backing on the tape can flake off if it is not firmly bonded to the base material. If the loose particles are deposited in the transport mechanism, they can clog the video head and scratch the tape.

Signal Degradation

Several electronically related defects can degrade the audio and video signals if the VTR is out of adjustment. One of the more obvious is dropout. A dropout is a brief missing piece of audio or video information that is seen as a black or white flash in the picture or is heard as a short interruption of sound during the playback of an audio signal. It can result from dust in the air, a worn video scanner or audio head, or a minute lack of oxide on the tape that prevents it from recording momentarily. This type of dropout sometimes can be corrected by rerecording the spot in question. If the dropout has been copied from another videotape, however, it might not be correctable.

Another type of dropout can occur if the sensitivity adjustment on the VTR's dropout compensator is set too low. This problem can be corrected by properly adjusting the playback VTR and rerecording the area. Dropouts also might be caused by a manufacturing defect in the videotape.

Audio and video levels that are set too high or too low are two other common causes of signal degradation. Higher than acceptable audio and video levels can result in a distorted signal. Lower than acceptable levels can make the picture dark and grainy and make the audio unintelligible and full of hum.

The signal-to-noise ratio of videotape refers to the relationship between the picture content and the amount of noise, or graininess. This ratio can be affected by the quality of the oxide formula applied to the Mylar base. It is nearly impossible to maintain the same quality from one batch of videotape to the next, but a roll of videotape should be rejected if the signal-to-noise ratio does not meet the minimum standards of the tape or VTR manufacturer.

Signal-to-noise problems also can be caused by worn audio or video heads. The heads should be checked and, if necessary, replaced before blaming the videotape stock. Using the VTR alignment tape discussed later in this chapter will usually identify a worn video scanner or audio head.

Improper Reel Size and Type

The size of the reel used on one-inch VTRs has a lot to do with how the tape is handled by the transport mechanism. Although nearly all takeup reels used on one-inch VTRs are made of metal, plastic, or glass, the reels used on supply-side VTRs can be made of lightweight plastic. For example, on older one-inch VTRs, a six-inch or eight-inch plastic supply or takeup reel might not play properly if there is not enough mass in the plastic supply reel to create the proper tension during fast forward and rewind.

The most important thing is to use the right reel for the job. Newer VTRs such as the Sony BVH-2000 and the Ampex VPR-3 can handle odd-size reels on both hubs without damaging either the tape or the VTR, but using small, lightweight plastic reels on older models can result in chattering and damage or deform the tape if the torque is too great. Excessive tension can stretch or even snap the tape, cause the picture to lose tracking, or break up the signal during playback.

Therefore, when small plastic reels are used on older VTRs, a special metal flange designed to add mass to the supply reel should be clamped to the outside of the reel after it has been mounted on the VTR. This provides sufficient weight and mass for any size reel used on a one-inch VTR.

Poor VTR Maintenance

Keep the VTR Clean. Many of the technical problems attributed to the tape stock are caused by a dirty VTR or VCR. Just because you cannot see any dirt or oxide in the transport path doesn't mean there isn't any. Procedures for the proper cleaning of the VTR or VCR are included in the operator's manual supplied with the machine.

To ensure optimum picture and sound quality, it is recommended that one-inch reel-to-reel VTRs be cleaned with every reel change. Three-quarter-inch, half-inch, and quarter-inch VCRs should be cleaned at least once a day if they are being used in an off-line environment and more frequently (every other reel change) if they are being used in a mastering configuration, where picture and sound quality must be maintained. Alcohol is a popular nontoxic cleaner used by many facilities. Cleanliness also applies to consumer VCRs and ATRs. If you notice that the picture is getting snowy or grainy or the audio is muffled, it might be time to clean the audio or video heads.

Use the Alignment Tape. To confirm the operating condition of the VTR, you should use the manufacturer's alignment tape at least once a day. This verifies that the equipment is in working order and that the signal path meets the minimum standards set by the manufacturer.

The alignment tape might reveal video and audio head wear. If proper maintenance is performed to correct any defects, no serious signal degradation will occur in subsequent recordings.

Check for Wear and Tear. If the video heads are worn or are repeatedly getting clogged with loose oxide, it is time to replace them. Otherwise, they will continue to cause problems and will ultimately result in more costly and time-consuming repairs.

On reel-to-reel VTRs, make sure that the reel hub lock is firmly tightened. Otherwise, the reel might spin on the hub and cause mistracking or possibly tape damage due to stretching or cinching. Use locking hubs on vertically mounted VTRs to prevent reels from accidentally falling off.

On VCRs, be sure the cassette carriage is in good shape and is properly aligned. In addition, make sure that the cassette is inserted carefully into the VCR and that the tape withdrawal assembly pulls the tape out of the cassette without wrinkling.

Another source of problems is worn or out-of-alignment pinch rollers. If the rollers become glazed, they cannot grab the tape properly, resulting in slippage during both recording and playback. Never use Freon to clean rubber pinch rollers. Use alcohol sparingly to keep the roller surface dull enough to maintain tape contact. Pinch rollers that are not aligned with the capstan shaft will almost always wrinkle the tape and make it unusable.

Accidentally dropping a roll of videotape on the floor can nick, dent, or warp the rim. Always replace a damaged tape reel with a good one, as bent or nicked reels can tear tape, cause it to run off the guides, or even result in severe damage to the videotape or VTR.

Improper Handling

When you are finished using a tape, always fasten the end of the reel so that it will not flop around inside the case. Never use Scotch tape to fasten the tape end. Use special adhesive tape designed not to leave a sticky residue that might adhere to the video and audio heads and either damage or clog them.

Never use any type of masking or electrical tape to repair or secure the end of a videotape. During rewind, the tape can wind itself through the transport, pass over the audio stack, and even get caught around the spinning video head, possibly causing permanent (and expensive) damage to the video drum assembly.

Always trim worn or wrinkled tape ends so they do not damage the video heads as they wind through the transport. One-inch and smaller format videotape should never be spliced if torn or broken unless you have no choice. If the original master is torn, you might have no choice but to splice the tape. If an edited master is torn, you should reedit or copy those sections to another tape and repair the program with electronic editing.

If the ends are torn cleanly, you might try to splice the videotape using only specially designed splicing tape. If the ends are jagged, trim them evenly across the width of the tape. To create surfaces that will butt together, overlap the ends and then trim them.

Helical-scan recorders do not accept mechanical tape splices well, and even though you might splice a tape perfectly, you will always see some picture disturbance. Recording over a spliced area will not correct the problem because the extra thickness of the splicing tape will always cause some visible disturbance.

Before recording on or playing back a videotape, wind it from beginning to end, then rewind it to pack the tape evenly so it will not cinch or snap when the VTR goes into full wind or rewind. Using loosely wound videotape on any size reel can cause severe damage to the tape.

When a tape is returned to you, always check the box for notes that might indicate tape damage or some other defect that should be corrected before using the tape again. Also look for water damage, which will make the tape unusable in most cases.

Always store videotapes on end (vertically), not on top of each other (horizontally), especially if they are stored in cardboard boxes instead of heavy-duty plastic cases.

Environmental Problems

Humidity. If the ambient humidity in the tape storage area and around the VTR is excessive (above 60 percent), the tape oxide has a tendency to absorb moisture, much like a sponge. When this happens, friction builds up and a form of stiction might occur. Excessive moisture will contribute to head wear because the added moisture will swell the tape slightly and cause more tension on the tape as it passes by the video head.

Problems related to humidity can be reduced or eliminated by keeping the humidity between 40 and 60 percent and the ambient room temperature between 68 and 72 degrees. It is helpful to determine the optimum operating temperature and humidity range in your area. This information is available from most manufacturers in pamphlet form.

Dirt. Dirt, especially dust, is usually a more severe environmental problem than humidity. It is the single most common cause of excessive head wear. The editing and tape rooms should be kept clean and dust-free. The rooms should have no open doors leading outdoors through which the wind can blow fine dirt. The exterior racks and cabinets of the VTRs and associated equipment should be vacuumed and wiped clean with a damp cloth at least once a month to prevent dust buildup. If you have a computer floor—a "false floor" beneath the walking surface where computer and electronic equipment cables lie—the panels should be pulled up at least twice a year and the floor underneath the cables vacuumed.

This type of preventive maintenance is required if you want consistently high-quality operation. Remember, it is less costly to keep the VTR and the room clean than it is to spend money replacing expensive video and audio heads.

Smokers should never be allowed in the editing or tape room. Smoke is made up of tiny ash particles that can cause severe video dropouts and excessive video head wear. If people smoke around floppy or hard disks, head crashes can occur. A head crash is a failure of the magnetic head or disk surface due to smoke particles coming in contact with it. Such crashes can be very costly, not only in lost disks and heads but, more importantly, in the loss of irretrievable data.

Always wash your hands before cleaning, threading, or operating a VTR to prevent oil and dirt from contaminating the transport surfaces. Dropouts will occur if grease or body oil contaminates the videotape.

CONCLUSION

The problems outlined in this chapter are some of the more common ones you might encounter in day-to-day videotape operations. If you encounter a problem, don't jump to conclusions without calling for a full investigation by a trained technician. Although videotape is manufactured according to very high standards, care must be taken during all phases of production and post-production to maintain this high quality. You can never be too careful when it comes to videotape.

The Marriage of Motion Picture Film and Videotape

10

Although a great deal of television programming is currently produced solely on videotape, film is still a very important part of the television industry. This chapter describes the production and post-production processes of both film and videotape and explains how the two are blended to create broadcast programming.

SINGLE CAMERA VERSUS MULTIPLE CAMERA PRODUCTION

The two production and post-production methods used in creating television programs are the single camera, or film-style, technique and the multiple camera technique. The use of a single camera results in many takes of a single scene, while the use of multiple cameras results in only one or two takes of a scene with source material from several different cameras. The production technique, as well as the medium chosen (videotape or film), makes a big difference in the post-production process.

Generally speaking, single camera photography, whether on film or tape, is used to record action or adventure shows outside a controlled studio environment and for other productions in which location and other factors make the use of several camera setups infeasible. Multiple cameras are used extensively to shoot in-studio productions such as situation comedies (sitcoms).

As a general rule, the use of multiple cameras speeds up the post-production process but increases the production budget because of the need for extra cameras, tape machines, personnel, and videotape or film stock. In addition, a camera person is required for each film or video camera used on the set.

A multiple camera film production generally requires the same number of cameras as a multiple camera videotape production. Multiple cameras are used in two ways in a videotape production. The first way is to send the picture from each camera to its own VTR. In this sense, each video camera is "slaved" to one VTR, and everything this camera sees is recorded on videotape. This is known as a dedicated camera/recorder connection and is similar to multiple camera film production, in which each film camera records everything that the lens sees directly on film. (See Figure 10.1.)

The other way multiple cameras are used in video production is to feed all the cameras' output through a video switcher. (See Figure 10.2.) This allows the director to edit on the fly as the program progresses and thus reduces the number of edits required during post-production. One drawback of this is that once a camera switch has been recorded on videotape, this edit point cannot be easily altered unless other material has been recorded on a separate isolated camera from a different angle or the material is rerecorded as a pickup take later in the production.

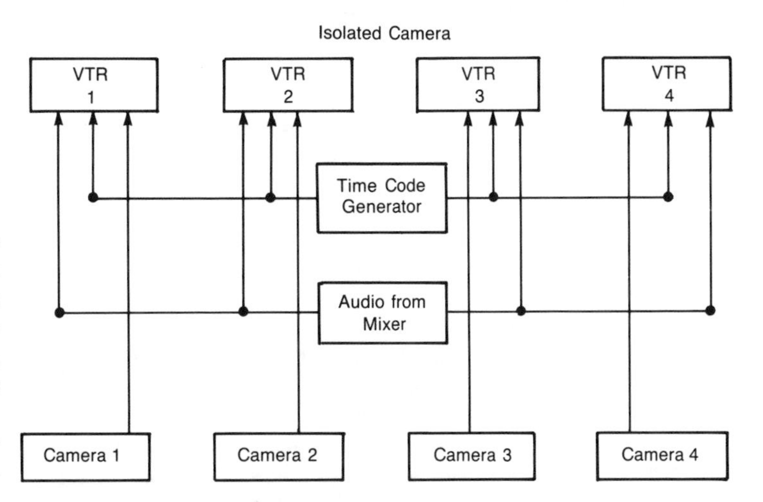

FIGURE 10.1
A block diagram of dedicated camera-to-recorder connections in a multicamera production. Each camera is isolated, allowing a wider selection of camera angles for editing. All the VTRs get the same time code and audio feeds, but each receives the video output from a single dedicated camera.

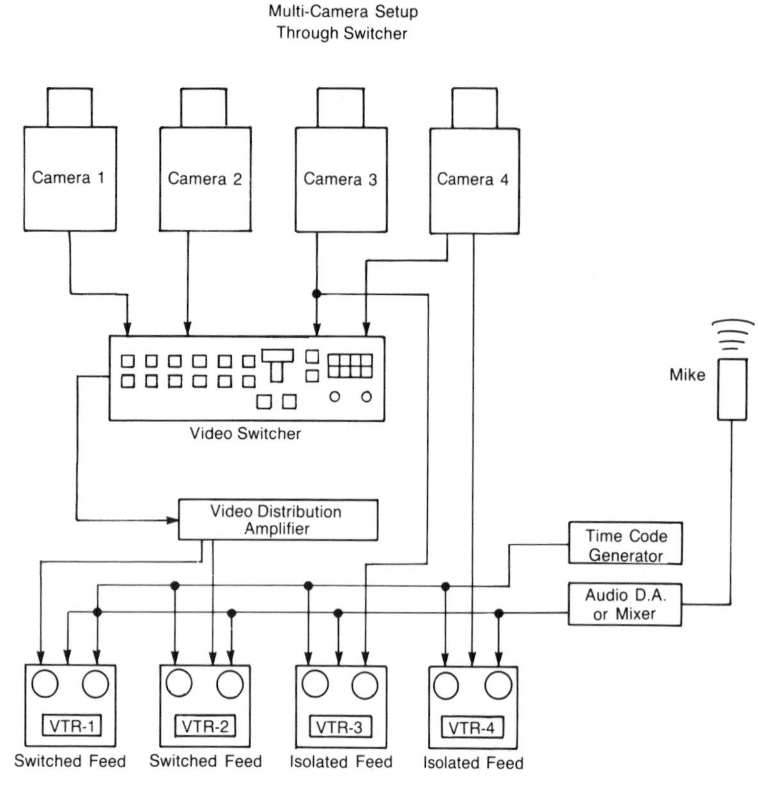

FIGURE 10.2
A block diagram of a multiple camera production using a video switcher to change cameras (or camera angles) while shooting, thereby allowing the director to edit on the fly. In this example, all the VTRs receive the same time code and audio feeds. The first two VTRs receive a picture feed selected by using the video switcher. VTR 3 and VTR 4 are directly connected to cameras 3 and 4, respectively.

One advantage of multiple camera production is the ability to more easily match action from one camera angle or take to another during editing because all the cameras are shooting the same action from different points of view. Another advantage is that fewer takes are required for a given scene. Since several cameras are photographing the same action simultaneously from several angles, the director has many options from which to choose. Although more costly in terms of equipment and labor, reduced production and editing time might offset these expenses.

A single camera production requires only one camera person and one or more assistant camera persons, who move the camera to a new location for each new setup. Those used to working in the film industry often prefer this type of production because they are familiar with it. The technique is the same for either film or videotape production.

The disadvantage of the single camera technique is that the cast must repeat the same action and dialogue over and over for as many takes and different angles as the director feels he or she needs to get the scene on film or tape. Sometimes the talent does not repeat the required action or dialogue exactly the same way each time, which can make editing difficult.

To simplify the discussion in this chapter, I refer only to the single camera, film-style production technique, as a description of multiple cameras might confuse the issue. For the most part, however, the techniques and terminology apply to multiple camera production as well.

THE POST-PRODUCTION PROCESS

This discussion centers on the post-production processes currently used to finish film for a film release and film for a tape release. I use a dramatic episodic television series as an example of these processes, since this type of show requires the most services to create a finished product and typifies the way all television programs are post-produced. To clarify the processes and their interrelationship, I discuss each step used in film and the corresponding step, if any, used in videotape, or electronic, post-production.

Developing and Printing the Film

In both cases, the picture image is photographed on motion picture film and developed to produce what is known as an action negative. In the film process, the action negative is printed by exposing the developed original camera negative in a printer against the print stock. This print is then developed, which results in a positive picture of the action. This positive picture is eventually used by the film editor in the editing process.

In the electronic process, the developed negative is not printed but is transferred in its negative form to a master one-inch videotape on a device known as a telecine machine, referred to as a flying spot scanner. Its purpose is to electronically convert the negative image into a positive color image and record this information on videotape, generally the one-inch type C format used in most of the United States, Canada, Mexico, and many other countries around the world. Newer videotape formats such as Betacam, the MII format, and the D-1 and D-2 digital formats also are used to record images for broadcast-quality pictures and might soon replace the established one-inch format as the industry standard. Some of these alternative formats are discussed in Chapter 13.

Flying spot scanners such as the Rank (Figure 10.3) are continuous motion devices. That is, they do not pull down the film one frame at a time with an intermittent jerk 24 times a second as most other projectors do. Instead, the film passes through the gate in a continuous motion, where it is scanned by an electron beam that converts the photographic image into an electronic television signal. Thus the camera original negative is not subjected to the hazards of rapid movement. The film is handled at least as carefully by this type of device as by high-speed laboratory film printers.

Figure 10.4 shows another popular flying spot scanner telecine device called the Bosch FDL-60 Digital CCD Telecine. CCD is an acronym for charged coupled device, which is a solid-state image sensor that converts the photographic film image into an electrical image.

FIGURE 10.3
The Rank Cintel MKIIIC flying spot telecine machine. (Courtesy Rank Cintel)

Recording Production Sound

The original production sound is recorded on a quarter-inch ATR designed specifically for motion picture or videotape production. The most popular brand of recorder is the Nagra IV series, which has been used for decades. Although other brands also are used for this purpose, I will use the Nagra as an example here. In addition to the production sound, the Nagra records the sync pulse, a signal that acts like electronic sprocket holes during the film-to-tape transfer process to maintain accurate tape playback speed, which is essential for proper lip synchronization.

In film, selected sound takes, also known as circled takes, are transferred to 35mm film. The film used to record the sound from the quarter-inch audiotape contains a single stripe of magnetic iron oxide along one edge similar in composition to the oxide used on ordinary audiotape. The sound on this roll of film should be in frame-accurate synchronization with the action negative and print.

Transferring the audio to videotape can be done in one of several ways. Some post-production facilities synchronize this 35mm magnetic sound track with the action negative and transfer both to videotape. At this point, the sound has already been copied down one generation, and if it is not handled carefully throughout the post-production process, it might end up being somewhat distorted on the final sound track.

To minimize this problem, the predominant practice is to transfer the sound directly from the quarter-inch production track to the one-

FIGURE 10.4
A Bosch Digital CCD telecine machine. (Courtesy Robert Bosch)

inch videotape master transfer roll. Another electronic alternative that greatly reduces the chances of audio distortion is to transfer the sound and its recorded sync pulse from the quarter-inch original to a three-quarter-inch digital audiocassette. A predetermined SMPTE time code also is transferred to the audiocassette and the one-inch videotape. This aids in the synchronization of the action negative.

The second alternative appears to be the most attractive. The transfer is effectively nondegrading as opposed to 16mm or 35mm magnetic transfers, which reduce the sound quality slightly from one generation to another. Although most sound recorded in motion picture and television production is quite good even with this slight degradation, with the second alternative, the use of digital audio eliminates most of these problems altogether.

Assembling Print Takes and Coding Film

If the material is to be edited on film using conventional film editing equipment, the negative is broken apart and only the print takes are assembled on a roll for printing. The transferred circled sound takes are generally in the same order as the corresponding picture roll and both are sent to the assistant film editor intact. This spliced roll of good takes is called the A negative roll, while all the unprinted outtakes are assembled on the B negative roll. The A negative roll is sent to the laboratory for printing, while the B negative roll is stored in a vault.

When the assistant film editor receives the 35mm magnetic sound track print takes, he or she synchronizes the picture and sound by using a visual aid such as a slate with a clap stick on the picture and the corresponding sound the clap stick makes on the sound track. A voice on the sound track gives the scene a take number (in effect, this is an audio slate), while the same visual information is seen on the slate photographed by the camera.

By running the sound track over a magnetic pickup head to locate the frame where the edges of the clap stick come together, the assistant editor identifies the frame where the clap first occurs. The editor then identifies the video frame where the clap stick's edges come together. The assistant editor marks these points on the sound track and picture with a grease pencil and lines them up in a film synchronizer. Once the action and track have been synchronized scene by scene and take by take, the material is ready to be viewed.

After synchronizing a roll of scenes and takes (about one thousand feet per roll), the assistant editor must provide a method of indexing the two strips of film to help maintain synchronization once the film is cut apart for editing. This is called coding. In a simplified form, coding is the process of printing a corresponding series of letters and numbers on the edges of the picture and sound track film.

The easiest way to do this is to line up the synchronized picture and sound in a sync block. This device is composed of two or more metal wheels with sprocket teeth. The wheels are exactly one foot in circumference and are positioned side by side on a common shaft. A large knob at the end of the shaft allows the user to rotate the wheels together either forward or backward.

The sound track is mounted on the sprocket teeth of one wheel, and the picture is mounted on the adjacent wheel. A short film leader is attached to the head of each roll, and the wheels are rotated backward so that several feet of both films are unwound. A permanent reference or X is made at the same point on each film. The two films are then removed from the sync block, and one is placed on a coding machine, or inker, which simply prints code numbers on the film. An adjustable numbering block can be preset to print any character combination at 1-foot increments. In most cases, a letter code identifies the roll number and remains constant throughout. An accompanying number code increases by one as each foot of film advances.

An identical set of letters and numbers is printed on both rolls of film. It is important to code both rolls before cutting them apart. By lining up the same code numbers on the picture and sound track, the editor will be able to maintain lip sync.

The code numbers are recorded in a special book along with other pertinent information. If the editor locates a piece of film and cannot determine where it came from, the code book will give the editor the complete history of that piece of film.

The Film-to-Tape Transfer

If the film is being transferred to videotape, the developed film negative is ultrasonically cleaned and inspected. Next, the negative is placed on a telecine machine. In the electronic process, the procedure of physically cutting out the unwanted takes on the negative is not necessary. When the roll is mounted on the telecine machine, any take on the reel can be found by shuttling through the roll at speeds up to ten times running time. Only the good takes are transferred to the one-inch videotape master, thus editing out the bad ones.

The negative color image is electronically reversed so that a positive image appears on the videotape, allowing the telecine operator to establish the proper density and color balance for each scene. The electronic controls on the telecine machine allow the operator to independently vary the intensity and color of the light, dark, and midrange portions of the picture. The objective of this telecine process is to reproduce the film accurately while maintaining the same dynamic range, or black-to-white ratio, within the limits of the television system.

During the telecine transfer process, time code generated by the telecine machine and applied to the tape is synchronized with the random time code added to the digital audiocassette. Once this synchronization is established, the digital audio is transferred to the sound track of the videotape.

Once the film audio and video are transferred, copies of the one-inch tape are made in smaller formats to be used for editing and viewing purposes.

Pacific Video of Hollywood, California, uses a slightly different process of transferring audio. At the same time the film is being transferred to the videotape, time code from the master tape is being recorded on a second digital audiocassette. This is used instead of the random time code on the original audiocassette to sync the digital audio to the telecine. Once the film has been transferred to videotape, all references to picture and track will be from the master tape, so it makes sense to have a second copy of the digital audio with a time code that is the same as the code on the master tape. Using this time code, the digital soundtrack is transferred to the master videotape in analog form creating a composite master tape. Remember that audio copied in the digital domain does not degrade, so this process can be repeated as many times as required.

Compiling and Viewing the Dailies

At a typical post-production studio, the transfer process from quarter-inch audiotape to 35mm magnetic film or a digital audiocassette starts at about 9:00 p.m., with the production audio for that day's shoot being brought into the sound department or a sound services company. Both types of transfer continue through the night and early morning, terminating somewhere around 3:00 or 4:00 a.m.

The picture developing process also starts at about 9:00 p.m. and continues to 1:00 or 2:00 a.m. This allows the printing process to start at about 3:00 a.m. The transfer process from film negative to videotape starts at about 4:00 or 5:00 a.m.

Timing is quite important to the production staff, as the footage must be viewed as quickly as possible on the day after it is shot. This is where the word *dailies* comes from.

The director and several other members of the crew generally view the previous day's footage at about 11:30 a.m. This viewing gives them a chance to evaluate the performances and to check for technical problems such as microphones in the shot, unwanted shadows, and lighting and composition problems. If dailies are run too late in the day, a second day's footage might be shot with the same technical problem.

This was brought home to me when I was working on a feature film shot on location in Mexico. Because of the remoteness of the location, 10,000 feet of film were shot before any film was developed. This amounted to several days' production. When the film was finally processed, the negative was quite dense, indicating a heavy overexposure. The overexposure was so great that the film could not be printed.

It was soon discovered that an electric light used as a marker device in the camera for synchronization had been stuck in the on position, in effect ruining every frame of film shot with that camera. Since the cameraman was unable to send exposed film to the laboratory on a daily basis, more than a week went by before the crew was notified of the problem. All the footage had to be reshot.

A disadvantage of film dailies is that while they are being viewed, the editor and assistant editor must wait until the end of the screening before breaking the viewing reels into individual scenes and takes. In videotape, copies of the dailies are typically made on three-quarter-inch videocassettes, which are distributed to the studio, the editor, and other interested staff. Thus the editor can begin work as soon as the material has been transferred.

Editing

While the film editor uses the synchronized work print and 35mm magnetic sound track, the video editor uses one of the videotape copies of the film transfer. In the film process, the original camera negative and quarter-inch audiotape are stored in the vault. Similarly, in video the one-inch master tape and digital audiocassette are placed in the vault until they are needed for the final conforming process.

Using conventional film editing equipment such as the Steenbeck, Kem, or Moviola, the film editor edits together a rough assembly by physically splicing picture film and magnetic sound track to create a work print that follows the continuity of the script. The end result is two reels, a cut work picture and a cut sound track.

In the rough cut, or first assembly, the film is usually edited long. This first edit shows basic angles but does not always attempt to make tight cuts. Based on feedback from the director and other key people, the editor reedits the work print, cutting the program as tightly as possible to bring it in line for television release. This version, called the editor's cut, is still a little long, as the editor knows the director and producer will want to make additional changes. The network also might send a representative to view the program, and he or she might suggest even more changes. When everyone is satisfied or time runs out, whichever comes first, the work print is referred to as the final cut. At this point, the reels are locked and no further changes can be made.

Videotape editing systems can be divided into two distinct groups: linear and nonlinear. (See Chapter 8 for a discussion of how the two types of systems operate.) Linear editing systems are very fast and efficient for creating a first cut but poor for making changes. Nonlinear systems, which edit by playing back several identical copies of the dailies through a switcher, were developed to facilitate the process of changing and reediting tape. Although nonlinear editing systems do perform this function well, some types of programs, such as multicamera productions, are better served by linear editing systems.

When all is said and done, the film editor ends up with a cut work picture and a cut sound track, and the videotape editor produces an edited work tape, if desired, and an EDL (see Chapter 8).

Creating Titles and Optical Effects

After the program has been edited and approved, the titles and optical effects must be created. In a film program, this process might take place while the work print is being edited, depending on the schedule. Titles, opticals, and other types of special effects are usually created in a separate facility.

After the titles and opticals are completed, the edited film work print is generally turned over to a negative cutter. A set of key numbers, which were photographically printed on the film stock by the film manufacturer and appear only when the film is developed, appear once every foot along the edge of the 35mm film. These numbers appear on the film negative and on the work print, but they do not bear any relationship to the code numbers deliberately printed on the film. Key numbers are used by the negative cutter to conform the original camera negative to the edited work print.

Prior to the start of the film conforming process, the negative cutter makes a written log of every key number in the work print, cut by cut. The cutter then goes into the film vault and uses the key numbers to select the rolls of camera negatives that are required to conform a given reel of film.

The cutter must locate and match cuts to the work print frame for frame. He or she uses scissors to cut the camera negative in the middle of the frame on either side of the actual cut. All the cuts are taped together on a reel, which is sent to the laboratory to be spliced.

In the electronic process, the editor uses the EDL generated by a linear or nonlinear editing system to conform the one-inch videotape in a linear fashion, creating a second-generation edited master tape. The edited master can be assembled sequentially or by a process called B mode assembly.

Sequential assembly means that the second-generation master is assembled starting with the first edit in the list and continuing to the end of the program. As each new edit is brought up by the computer, it checks to see whether the correct reel number is mounted on any of the playback VTRs. If it is, the system automatically makes the edit, joining the current and previous edits back to back. If a reel is not already mounted on one of the VTRs, the operator will have to unload one of the source reels and substitute the source reel called for by the editing system. This slows up the assembly process.

B mode assembly allows the system to make edits relative to their positions on the master tape, leaving blank spaces for those edits not yet recorded. For convenience, the operator usually loads the source reels in the order they were originally shot or transferred from film. The editing program selects and records all the material from the first reel that will be used in the program, positioning each edit frame accurately on the second-generation master. When the system has transferred all the cuts from the first reel to the master tape, the operator mounts the next reel, continuing until all the edits have been transferred. The primary advantage of this type of assembly is the reduced number of reel changes.

For the B mode process to work efficiently, the EDL loaded into the computer must be error-free or these errors will appear in the master. EDLs generated by nonlinear video editing systems are clean—that is, they do not contain any errors because each edit is based on stored time code information. If a change is made at any time during the editing process, the EDL is automatically corrected. This is not the case with linear video editing systems, which store all edits, including those no longer wanted. Unless the EDL is cleaned before on-line assembly, those unwanted edits will cause problems. (See Chapter 8 for the details of cleaning lists.)

Although the resulting edited master contains sound, a new digital master audiotape is created using the same EDL used to create the auto-assembled picture. This ensures that sound quality will be maintained through the sound mixing and compositing stage. The three-quarter-inch digital audiocassettes made at the same time the film was transferred to one-inch videotape are used to conform the digital master audiotape. This edited digital master audiotape has the same set of time code information that was used to conform the picture and is thus in exact frame sync with the edited video master. Because of the very high quality of digital audio, creating this new digital audio master is in essence the same as if the editor were to go back to the quarter-inch production recording tape and physically splice the edits together.

Optical effects such as dissolves, most wipes, and fades are part of the normal electronic assembly process, so there is usually no additional charge to create these effects. To generate titles and credits in the electronic domain, it takes a skilled operator about an hour to type in all the names, position them, and add color or drop shadows if desired. It takes another half hour for the operator to add them to background material already in the program.

The electronic counterpart of the film cut action negative and cut sound track is a one-inch edited master videotape that contains both action and sound and, in some cases, a separate master digital audiotape.

The electronic process has several advantages over the film editing process. It is faster, and you can try an edit any number of ways without cutting a frame of film. Producing one episode of a one-hour dramatic series on film, assuming a standard three-week turnaround, requires three editing teams, each including an editor, assistant editor, and one or more apprentice editors. In contrast, a one-hour episode produced on a nonlinear videotape editing system requires only two editing teams. Thus, even though nonlinear videotape editing systems cost a lot more (about $150,000 each) than film editing machines ($12,000 to $30,000 each), the amount of money saved by having to hire only two editing teams is just about equal to the additional cost of a nonlinear system.

Off-line linear editing systems generally use only one editor without an assistant, although sometimes an assistant is used to keep track of the material. Even though the off-line linear process requires fewer people, it is much slower and less efficient than the nonlinear process.

Timing

The next step in the film process is timing, which refers to analyzing each cut relative to color and density and establishing a set of corrections that can be applied to the negative during the printing process. The result of this process is a timed answer, or first trial, print.

A similar process, called electronic timing or tape-to-tape color correction, occurs in electronic editing. To understand this part of the electronic process, let's review the printing of dailies in film. The purpose of making a daily film print is to feed back to the production staff technical information relating to the image on the camera negative. Since the daily print is used only by the film editor to build program continuity and by the negative cutter to match the negative when conforming, the color and density at this stage are not very important.

A different approach is used in the electronic transfer procedure. The telecine operator establishes a series of correction values so that the production staff has a reference for exposure and other technical elements. The videotape master will be used as a basis for all subsequent corrections, up to and including the final edited master used for broadcast or syndication. Therefore, care must be taken in the film-to-tape transfer.

During auto-assembly, minor density and color corrections might be made, but a complete timing of each scene is not done. Scene-by-scene color correction density and contrast values are applied to the assembled master later, producing a timed third-generation master.

Refining the Sound

In the past, very little effort was put into creating a top-quality sound track for a television program, even though enormous effort might have been put into creating the picture portion of the program. Digital audio and stereo broadcasting have made giant strides toward providing viewers with quality sound.

Let's first discuss the preparation of the sound track used in the film process. Under conventional procedures, film reels are built in 10-minute segments, mostly for convenience. Raw film stock usually comes in 10-minute lengths, and larger reels can become quite heavy and unwieldy. A 1-hour film program is generally stored on five 10-minute reels, since the program content is only about 52 minutes.

Each of these 10-minute reels must be prepared for the sound mixing process. To do this, reels containing sound effects, music, or additional dialogue are synchronized with the picture. The production sound track is then labeled the A track, while additional tracks are labeled B, C, D, and so on. The sounds are put on different tracks to make it easier for the sound mixer to blend these elements together. The greater the separation, the smoother the end result.

For example, let's say a reel of dialogue contained alternate interior and exterior takes with corresponding level and quality changes. The sound editor would split up these tracks by putting all the interior lines on one reel, called the B reel, and all the exterior lines on another reel, called the C reel. Each reel would have an appropriate spacer leader to maintain synchronization with the picture. Sound reprints also might be ordered so that when the B and C reels are constructed, a small amount of overlapped in-sync dialogue can be added to adjacent reels to soften the blend in a process similar to that used to create a picture dissolve.

In addition to dialogue lines recorded during production, there might be looped dialogue lines, which are recorded in a sound studio to replace part of the original dialogue. These new lines are recorded on one track of the 35mm three-stripe magnetic film and are later substituted for the original lines.

Additional sound effects, such as traffic noise, brakes squealing, and horns blowing, might be recorded on separate reels. Nearly all the sound effects used in television or feature films are supplemented by custom sound effects obtained from a library or created in a sound effects studio. Often a good original sound effect picked up during location shooting will be added to the post-production studio's own sound effects library for future use.

One type of live sound effect is called a "Foley," named for Jack Foley, the man who first created the effect. Basically, the edited work print is projected onto a motion picture screen, and sound effects specialists are brought into the studio to duplicate real-life sounds in sync with the picture. Foleys are used primarily for effects where exact synchronization is difficult to achieve, such as in creating footsteps.

Once all the sound effects, music, and additional dialogue have been constructed on separate reels, the sound mixer can set the level and equalization for each. This ensures that when the sounds are eventually mixed, they will blend together to produce a balanced sound track in synchronization with the picture.

All these audio elements end up on 35mm magnetic sprocketed film. A typical filmed 1-hour television program ready for the mixing, or dubbing, process might include 4 or 5 dialogue tracks, 10 to 15 sound effects tracks, and 2 or 3 music tracks. Thus each 10-minute reel of video could have more than 20 additional reels of audio.

During the mixing process, all these reels will be combined into three individual mixed sound tracks—mixed dialogue, mixed music, and mixed sound effects. The resulting magnetic track is a 35mm three-stripe. The three mixed elements on the three-stripe are later combined to generate a composite of the tracks, which is what eventually becomes the composite sound track for the picture. The reason for not combining the three tracks immediately is that if the film is sold to a foreign market, it will be necessary to dub only a new voice track. The sound and music tracks can remain the same.

The next step is to combine all three tracks into one final composite while transferring them to a 35mm optical sound track negative. After this negative sound track is developed, it is synchronized with the cut picture negative, and a positive answer print is made. If the answer print is approved without changes, more prints are made to meet the network's requirements. If changes are required, the individual reels are reprinted until the client is satisfied.

In the electronic process, all the preparatory work for creating the sound track is essentially the same, except instead of building multiple sound element reels for each ten-minute roll of picture, the sound elements are recorded on a multitrack recorder, which typically accommodates 24 tracks on a two-inch audiotape. A three-quarter-inch or other small format videotape copy of the edited master picture containing burned-in time code numbers and running in synchronization with the sound track serves as a frame-accurate picture guide for the sound mixer.

The mixing session is conducted the same way as in film, with 3 channels open on the multitrack recorder to record the 3 mixed tracks. If a mixing session requires more than the 24 tracks available on one recorder, additional recorders can be connected in sync with the first to give the mixer and editor unlimited flexibility in the number of tracks available for the rerecording process. Instead of using the 3 mixed tracks to make an optical sound track negative for printing, however, the operator mixes them together and transfers them to the timed edited master as a composite mix in a process called the layback. This composite replaces the original edited but untreated production sound track.

If many copies of a filmed show are required for syndication, duplicate negatives and sound tracks will be made from the cut picture negative and 35mm three-stripe in either the 35mm or 16mm format, depending on the needs of the syndicator. If many copies of a program on tape are required in different formats, they can be made simultaneously by putting the master tape on a playback VTR and placing as many rolls of videotape stock on as many recorders as are required. If three or more identical masters are required, they can be recorded during the scene-by-scene timing process. When the composite sound track is laid back, it can be put on all the masters at the same time.

CONCLUSION

Today a large percentage of television programs formerly finished and released on film are now being shot on film and edited and finished on videotape. Since television in its present form is not capable of generating the high resolution needed to take full advantage of film, nothing is lost by transferring film to videotape. The "film look" is preserved during the transfer process and is maintained throughout the post-production process. In addition, using videotape for post-production can save the producer money. In fact, the money saved is sometimes the difference between being able to produce a program and not being able to produce it.

Using electronic post-production for feature films is not yet cost-effective. All the elements of feature films must be created and finished on film to preserve the high quality necessary for large-screen theatrical projection. For the time being, electronic post-production is effective only in the television industry.

The Film-to-Tape Transfer Process

11

Transferring motion picture film to videotape is an integral part of the post-production process. In Chapter 10, I discussed how the film-to-tape process works. In this chapter, I reiterate the importance of using this process and discuss some problems that can result from it.

WHY THIS PROCESS IS IMPORTANT

Although the majority of television programs are still shot on motion picture film, most are generally finished on videotape. Not only is this cost- and time-efficient, but also the look of film is retained when the material is transferred to videotape. In addition, the film editor's creativity seems to be enhanced when he or she edits on tape because the edit can be remade in several different ways without physically altering the tape. Each time a revision is made on film, the editor must cut the film, splice it, take it apart, add or remove trims, and resplice it. This procedure is time-consuming and frustrating, and it can damage the film.

In my discussions with a number of film editors, I have found that all see the value of videotape editing but most are not yet convinced that videotape editing systems are capable of performing the same functions as film systems. Nevertheless, once they have been exposed to the new breed of tape editing systems, most would not go back to editing film in the conventional manner.

TRANSFER SYSTEMS

Transfer systems are obviously the key to the film-to-tape conversion process. Although most transfer systems are known generically as telecine machines or film chains, they can be divided into two classes. The older type of machine consists of a special film projector and a tube-type television camera interfaced in such a manner that they act as one unit to transfer motion picture film to videotape at a fixed rate of 24 frames per second.

The second, more popular type of machine, known as a flying spot scanner, uses a single electron beam to scan the film rather than a conventional television camera pointed at a film projector. Electronic circuits in the unit convert the beam to video signals that can be either broadcast or recorded on videotape. Today most transfers are made using this type of equipment.

The advantages of the flying spot scanner are many, but one of the more notable is the ability to operate at speeds other than 24 frames per second. Another advantage is that the picture is advanced on a continuous basis as opposed to being intermittently pulled down as in conventional projectors. Flying spot scanners also handle original film negatives very well, and it has been shown that transfers using film negatives result in noticeably better quality than those using film prints. Finally, the flying spot scanner offers the advantage of being able to conduct high-speed searches without the operator having to unthread the film from the projection gate.

TRANSFERRING BLACK-AND-WHITE FILM TO TAPE

When motion picture film was transferred to videotape for broadcast in the early days of black-and-white television, a 60-minute film program always played back on videotape in 60 minutes. This was because the videotape player was referenced to the 60 Hz per second power line frequency and black-and-white videotape ran at exactly 30 frames per second. Therefore, a one-for-one relationship would always be maintained between the running time of the film and the videotape transfer of that film.

As television and film technology progressed, the one-to-one relationship no longer existed. Standard motion picture film is currently photographed at 24 frames per second, which is known as sound speed. When film is transferred to videotape, it is played back on the projector at 24 frames per second, but it is recorded on videotape at 30 frames per second. This unequal conversion results in the final tape being electronically stretched so that its timed length is longer than that of the original film. In broadcasting, this discrepancy can be a problem. Sometimes a film can be transferred and broadcast as is, but in other cases, a film must be edited and timed after transfer to fit a given time slot.

Some time ago television engineers tried to find a way to convert 24 film frames to 30 video frames by recording certain film frames twice in the hope that doing so would add six extra frames of video information to each second of videotape. This technique resulted in an annoying jerkiness or flicker in moving images.

Finally, the three-two conversion process, also called the 3/2 pull down, was developed. This method results in flicker-free motion and a smooth transition from one frame to another.

To understand how the 3/2 process works, you must understand what makes up a television picture frame. In television, the frame rate is 30 picture frames per second. Each frame is made up of two fields, an odd field (called field 1) and an even field (called field 2).

A complete television frame contains 525 lines of information, and thus each field contains half the picture information (262½ lines). When both fields are scanned together, they produce one complete frame.

One second of time consists of 30 frames, or 60 fields. This fact allows a telecine transfer device to convert 24 film frames per second into 30 television frames per second.

The top half of Figure 11.1 represents four film frames, and the bottom half represents five television frames. During the transfer process, video frame 1 gets the image of film frame 1 in fields 1 and 2. Video frame 2 gets film frame 2 in fields 1 and 2. Video frame 3 gets film frame 2 in field 1 and film frame 3 in video field 2. Video frame 4 gets film frame 3 in field 1 and film frame 4 in field 2. Finally, video frame 5 gets film frame 4 in fields 1 and 2.

In other words, film frame 1 is scanned by the telecine for two video fields, film frame 2 is scanned for three video fields, film frame 3 is scanned for two video fields, and film frame 4 is scanned for three video fields. This 3/2 alternate scanning technique is how the system adds the six extra frames. In effect, the system adds an extra television frame for every 4 film frames (for a total of 5). If you multiply 4 film frames by 6, you get 24 film frames. If you multiply 5 television frames by 6, you get 30 video frames. Thus, at the end of one second of time, you have 24 film frames, equaling 30 video frames.

TRANSFERRING COLOR FILM TO TAPE

When color videotape was developed in 1957, it was accompanied by a multitude of technical problems, some of which still plague us today. A major problem was that, for various technical reasons, the video frame rate had to be slowed by exactly one tenth of 1 percent to

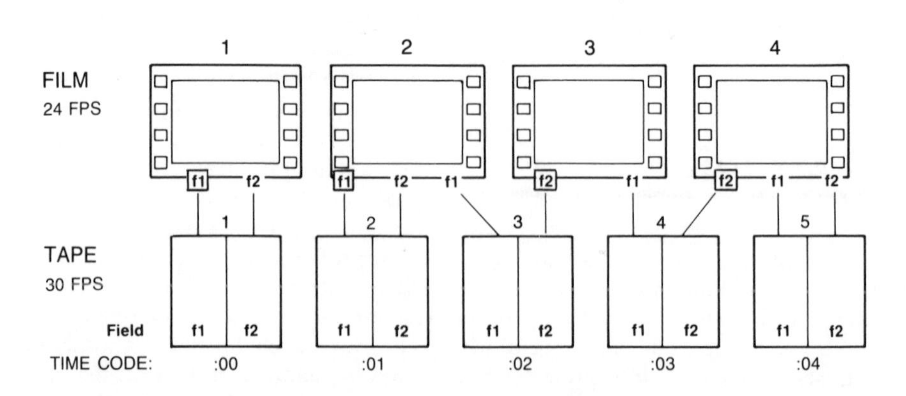

FIGURE 11.1
The 3/2 pull-down relationship for transferring film to videotape.

generate NTSC-compatible color on videotape. Thus, instead of recording and playing back color videotape at 30 frames per second, you must record and play it back at 29.97 frames per second. Because telecine transfer equipment must be referenced to the same video frame rate, the film transfer rate also has to change by one tenth of 1 percent to 23.98 frames per second. This slower rate creates a color film-to-tape frame time discrepancy. One second of material on film transfers to 1.03 seconds on videotape.

In other words, when 0.03 is added to every second for 60 minutes (0.03 seconds times 3,600 seconds per hour = 108 extra frames), the playing time for material transferred to color videotape is extended by 3 seconds and 18 frames every hour (108 frames divided by 30 frames per second). That's 108 frames added to the length of a one-hour program.

Even though 16mm and 35mm both run at 24 frames per second, the frame image size is quite different. The 35mm film runs at 90 feet per minute, while 16mm film runs at 36 feet per minute, a 2½-to-1 ratio. The 16mm film contains 40 frames per foot, while 35mm contains 16 frames per foot, again a 2½-to-1 ratio.

For a film program already finished on film, the only way to resolve this problem is to reduce the physical length of the film before transferring it by editing it to run 2 feet and 7 frames per hour less in 16mm (or 5 feet and 7 frames in 35mm) than it would if it were being projected on a motion picture screen. This is the same as a reduction of 0.1 percent in overall length. A proportional reduction of both film and tape can be described in the following way: tape—0.1% × 30 frames/second = 0.03; film—0.1% × 24 frames/second = 0.024. From this we can conclude that the reduction of both media will be equal, maintaining the 4:5 film:tape frame ratio.

The second problem concerning color transfers resulted from the use of time code. When time code was first used in an editing application, it was called color time code. In this form, it displayed the hours, minutes, seconds, and frames consecutively around the clock 24 hours a day. Today this is called non–drop frame time code.

If non–drop frame time code is used to time the length of a program on videotape, whether it is made from a film transfer or live video to videotape, the tape will play back 3 seconds and 18 frames longer than what the time code says it should. In terms of network broadcasting, the program running time would increase by 1.25 seconds every 20 minutes. This posed a severe problem for television broadcasters, who require accurately timed programs to plan commercial insertions and for other uses.

When using non–drop frame time code, broadcasters were forced to measure the length of every taped program in real time using a stopwatch to determine the exact commercial break positions within the body of a program instead of relying on time code information. Drop frame time code was developed to solve this problem. See Chapter 5 for a detailed discussion of this concept.

Today, most telecine machines are designed to operate at a fixed rate of 59.94 Hz per second in color, which still generates approximately 30 video frames per second. The telecine machine runs at approximately 24 film frames per second, which determines the speed at which the film is played back on the device.

It should be pointed out that even if a film was broadcast directly on the air, the actual running time of the film playing back on a flying spot telecine device would be the same as if it were first transferred to tape and then broadcast.

There are two separate issues here. The first is the 108-frame difference in the film-to-tape transfer process, and the second is the 108-frame difference between drop frame and non–drop frame time code. To avoid time problems when transferring film to color videotape, two rules must be followed. First, the editor must allow for this time expansion by cutting the total length of the film to compensate for the extra time generated by the film-to-tape transfer operation. The second is to apply drop frame time code to the videotape during the film-to-tape transfer operation so that a frame-accurate determination of the actual running time on videotape can be made.

Feature films can be transferred to videotape in an uncut form and timed after transfer to determine commercial positions within the body of the film. If the film is found to be too long for a given time slot, it can be edited on videotape to conform to the allotted time. In this case, the 59.94 Hz per second frame rate and the use of drop frame time code are of no significance, as editing the tape will compensate for time and length differences.

Once motion picture film is transferred to videotape and edited to program length, drop frame time code is used exclusively by the networks and local television stations to determine the placement of commercials and overall running time. It is expected that non–drop frame time code will be phased out altogether by the mid-1990s.

A Career in Videotape Editing

12

Videotape editing is one of the more creative aspects of the production and post-production process and is a controlling force that contributes to the success or failure of a project. Creative editing requires an ability to visualize an edit, remember how you want it to appear, and then make a good cut. A well-edited program should not look as though it has been edited.

In this chapter, I specifically refer to getting a job and working in the broadcast industry. Most of the information applies equally to other fields of communication that use videotape editing to generate programs.

ENGINEER OR ARTIST?

In the early days of videotape editing, many directors felt that the editor's primary role was that of an engineer or technician. Indeed, when videotape first appeared in 1956, most of the people who operated editing equipment were technicians who had been pressed into service because of a lack of people trained in the field of videotape editing. Unfortunately, only a handful of these technicians had any desire or the necessary training to become tape editors. Tape editors today still find it difficult to escape the stigma of being "only a technician." This stigma is due in part to the absence of an apprenticeship program for videotape editors. (The film editor unions require such an apprenticeship.)

In the early days of videotape, television facilities hired technical people with film backgrounds to splice tape under the supervision of the director or his designate. These engineer/operators were not really editors and had virtually no creative input into the project.

When kinescope recordings (monochrome motion picture film recordings of television images) came into use, the people who prepared the film for broadcast also were designated as engineers, since most of these people had little or no training in film-handling techniques. At that time, the term *editing* referred to repairing broken film, replacing or deleting commercials, and adding leaders when the film was ready to be sent to the laboratory to be printed for distribution to television stations across the country.

During the 1960s and 1970s, the role of videotape editing began to change at the networks and at independent production companies. The editor's role was expanded to include creative skills as well as mechanical know-how. Figure 12.1 shows the author editing and splicing two-inch videotape on "Rowan and Martin's Laugh-In" series. Directors preferred working with an editor who could take a script and put together a program without direct supervision of every cut. The editor was left alone to do a rough cut so the director could attend to his or her other affairs. The editor's role has continued to grow along these lines and is now one of the more important roles in the post-production process.

FILM VERSUS TAPE EDITING

The film editor has the responsibility of molding the material into a story, often without the director's constant guidance. The editor views the film dailies with the director, who selects

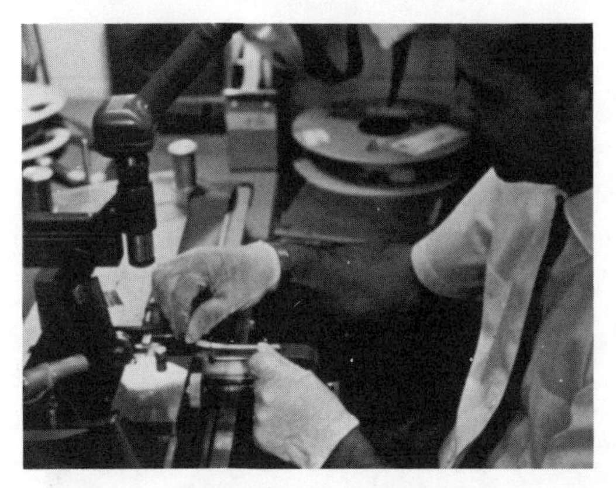

FIGURE 12.1
The author is shown splicing two-inch videotape for "Laugh-In."

the takes he or she wants to use and gives the editor any special instructions relating to the day's material. During this meeting, the editor and the director work as a team to choose the best material. When the director leaves, it is up to the editor to edit the film so that it represents the story outlined in the script and conveys the feeling or mood the director desires. Figure 12.2 shows the author some 15 years later performing electronic editing, following instructions given to him by the director.

After the editor makes the first cut, he or she meets with the director to review the material. The editor then recuts the work print, making the director's changes and perhaps adding a few of his or her own. The cooperative effort of the editor and director, as well as the editor's opportunity to work without supervision, is the basis of editorial creativity.

In contrast, as I have already pointed out, videotape editors have long been viewed as technicians who lack the creativity and storytelling skills necessary to play an integral role in the creative process. This perception is changing, and directors and producers are relying more on the skills and creativity of their editors. As more and more editors break free of the technician stereotype and the creative virtues of videotape become more widely accepted, the role of the videotape editor is sure to grow.

EDUCATION

Formal Education

A high school diploma and a basic knowledge of English is a must so that the editor will be able to understand training manuals and converse intelligently with his or her employer and clients. Communication skills are extremely important because the editor constantly deals with people on a one-to-one basis.

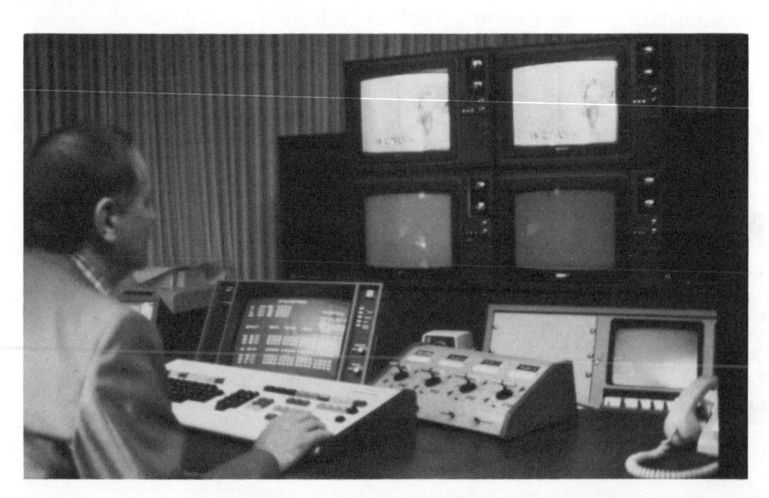

FIGURE 12.2
Here the author is doing electronic editing on an off-line system.

If a student takes basic computer courses, he or she will have a better understanding of the relationship of a computer to videotape editing and will be better able to get the most out of the editing. Even editing home movies on film or tape can give a person some understanding of how scenes relate to each other and what makes a pleasing edit.

If you have the opportunity to get formal training in film editing or tape editing, by all means do so. My training as a film editor made it relatively easy for me to move into videotape editing because I could apply most of the skills I learned as a film editor to videotape editing.

Although it is not required that an editor be able to repair the electronic equipment used in editing, he or she should not be intimidated by such equipment. Most editing facilities do not require an engineering background or degree unless the person is expected to do a highly technical job; however, the editor should be willing and able to learn how to make minor adjustments to the equipment.

Some colleges that offer electronics and computer courses, as well as classes in film and tape editing, allow you to take advanced courses without having to take any prerequisites, but these classes generally fill up fast. Although some schools do have fairly sophisticated editing equipment, most schools cannot afford to buy and maintain such equipment. Obviously, schools in areas such as Los Angeles and New York, where there are numerous video production companies, provide more classroom opportunities. In general, the best place to get formal technical training is at a trade school or college.

Commercial Training

The CMX Corporation of Sunnyvale, California, and the Convergence Corporation of Irvine, California, both offer training classes in the operation of their equipment. The classes are small, averaging about ten students per class, so the hands-on experience is relatively good. The cost is high, however, averaging between $500 and $900 per week plus expenses.

Most post-production facilities rarely offer training courses to anyone except their own employees. These facilities are in the business of making money, and training people is not a money-making proposition.

Internship Programs

Each year the Academy of Television Arts and Sciences awards a small number of internships in about fifteen areas of television production and post-production. One of the internships is in videotape editing. The student has an opportunity to observe at a post-production house for six weeks during the summer and to experience firsthand what it takes to put a television program on the air.

Although most union regulations prevent any hands-on operation of the equipment, the student is able to ask questions of the editor and, in most cases, the clients. For more information about the student internship program, write to the Academy of Television Arts and Sciences, 3500 West Olive Avenue, Suite 700, Burbank, CA 91505-4628.

The Academy of Motion Picture Arts and Sciences also has an established internship program. For information about this program, write to the Academy of Motion Picture Arts and Sciences, Beverly Hills, California.

Internships also might be available at local television or cable television stations or at the three major networks. It never hurts to inquire about internship possibilities, as an organization might agree to take on an intern even if no specific program exists.

GETTING YOUR FIRST JOB

Videotape editing is very competitive, as are most areas of the television industry, and any edge you can acquire will be of some value. If you are out of school, try to get a full-time or part-time job with a local television station or production facility, even if you are just a "gopher" or have to work in the mail room. The pay might not be great, but the opportunities to meet people will pay off.

One good way to meet people in the industry is to attend lectures, seminars, technical meetings, and conventions where industry people congregate. Joining organizations such as the Society of Motion Picture and Television Engineers will give you many opportunities to meet people and ask questions. If you are in school, most organizations offer a student membership, often at a reduced rate.

Many people think that videotape editing is a glamorous job, especially if it entails working in Hollywood. If there is any glamour, it's relatively rare, but one thing there is a lot of is hard work. Although you might make a name for yourself in the industry, success does not come overnight. You have to pay your dues, and it might be quite some time before you get a big break or your abilities are recognized.

I started my professional editing career at NBC in Hollywood as a temporary film editor. That job lasted more than 17 years. I was able to take advantage of a fortuitous situation, and it is important that you do the same. Even if a particular job is not exactly what you want to do, remember that a small break can open the door to your entire career.

Vacation Relief

Each year the three major networks hire temporary employees to replace permanent employees during the summer vacation period, which usually runs from the beginning of April to the end of October. Although applicants do not have to have any prior experience, the networks are partial to those who have had some training in college or at a trade school. In some cases, vacation relief employees are retained on a temporary or full-time basis after their initial assignment is over.

Learning on the Job

As I mentioned before, the best way to get started in videotape editing is to get any kind of job in a small facility and keep your eyes and ears open. You can learn a great deal about post-production and at the same time learn about other job opportunities. Do not make a pest of yourself when someone is in the middle of a complex operation. Watch and take notes, then ask questions when it is more appropriate.

It is very important not to let your enthusiasm to learn interfere with the job you were hired to do. Under certain conditions, you might be in a position to learn while you are doing your regular job. If this is not the case, wait until the day is over and, if your employer does not mind, stay late to observe an editing session in progress.

Underqualified people sometimes think they can do a higher-level job even though they do not have the necessary training or skills. Be honest with yourself and others when pursuing such a job. In preliminary interviews, do not tell the interviewer that you are capable of performing a certain task if you have never done it before. Remember that once you are in the editing room, technical skills are hard to fake.

Feel your way slowly, but keep your eyes and ears open. Your willingness to work, how you get along with other people, and your ability to give and take instructions and carry them out to the letter are all part of paying your dues.

My Hiring Requirements

I have only three requirements when I hire someone for an editing job. The first is that the applicant must have at least a high school education. The second involves the person's background and abilities. If the job is an entry-level position, the person must be able to learn the fundamentals of that job. In other words, the person will not be able to advance without mastering the job he or she is currently doing. Finally, I look for an individual who can get along with other people and can take constructive criticism. Although I admire someone with initiative, it is important not to appear pushy. That's the fastest way to alienate co-workers.

A TAPE EDITOR'S DUTIES

A videotape editor is in charge of the rest of the editing crew and is responsible for the technical quality of the edited videotape. The editor also is responsible for seeing that all the equipment needed for the job is in place, that all the crew members are present and ready to work, and that the edit base material upon which the program will be recorded has been prepared. In addition, the editor is responsible for making sure the client's tapes are in the editing room and that everything is organized so that the project will flow smoothly.

During and after an editing session, the editor is responsible for filling out the work order, which entails recording the time actually spent working on the project. Downtime should be deducted from the bill, and the cause and duration of the downtime should be noted. The amount of videotape used also must be logged and billed. If the stock used was supplied by the client, this must be indicated on the work order. All equipment used during the editing session but not included in the base hourly rate must be itemized and the amount of time it was used noted on the work order. Finally, the editor should make sure the client checks and signs the work order, signifying that he or she understands all the charges.

The Off-Line Editor

An off-line editor must be creative, have a good imagination, be able to visualize edits before making them, and have basic technical skills. This person must understand the use of time code and how it relates to editing. In addition, if an off-line editor expects to be proficient, he or she should thoroughly understand edit list management. Depending on the job and the client, an off-line editor might work alone or with others. I think that working alone is the most productive way to edit, as long as the editor is a self-starter and is capable of sticking to the schedule.

Most off-line editing systems use three-quarter-inch or half-inch VTRs and are relatively easy to operate. Newer off-line editing systems, such as the Ediflex, Montage, and Editdroid, do not use time code in the same way that conventional videotape editing systems do. Even though these systems are time code based, the editor does not see or deal with time code during an editing session. Instead, the editor edits by picture and sound the way film editors do.

The On-Line Editor

The on-line videotape editor must know more about television broadcast equipment than the off-line editor. An on-line editor usually starts out as a videotape operator, learning how to record and play back videotape on a variety of broadcast VTRs. An operator also learns how to make copies, which are called dupes or dubs, and how to make edit black, the master tape used to build the on-line edited master. The tape operator must be familiar with video character generators, know how to set up a color camera, and know how to connect one piece of equipment to another.

The next logical job to hold after tape operator is assistant editor. The assistant editor uses everything he or she has learned as a tape operator to assist the editor in an on-line assembly. The assistant editor must load tapes, anticipate the next reel change, and keep things moving swiftly and smoothly.

The assistant editor also keeps a close eye on

all the VTRs during the assembly to make sure the machines are operating at peak efficiency. While closely monitoring the record VTR, the assistant editor must alert the editor to any unusual glitches, flash frames, or other defects. It is much easier and cheaper to catch these problems as they happen than to go back and correct them later. From time to time, the assistant editor also will be asked to adjust audio and video levels. Finally, the assistant editor usually has the opportunity to learn editing skills and the basic operation of the editing system.

It is one short step from assistant editor to on-line editor. The on-line process is more technically oriented than the off-line process. In addition to being creative, the on-line editor must know how to use sophisticated equipment such as video switchers, digital video devices, character generators, and other tools not normally found in the off-line editing room. This technical expertise is one reason why on-line editors generally earn higher salaries than off-line editors.

Two Editors in One

Producers and directors often look for an editor who can handle both the off-line and the on-line sessions. Working with the program in the off-line session can make it easier for an editor to build the on-line master. In addition, if the off-line editor has to include notes in the edit list to remind the on-line editor of things to do during assembly, having one editor for both jobs saves time and cuts down on the possibility of confusion.

COMPENSATION

Salaries in videotape editing range from just about minimum wage to more than $2,000 a week. Added to these salaries are fringe benefits such as health insurance, pension plans, and dental insurance. Overtime, double-time, and triple-time pay also are available to editors who work more than an eight-hour day.

Free-lance editors usually receive a higher hourly rate than staff editors. Hourly rates for free-lance editors range from about $15 to $60 dollars an hour. Of course, free-lance editors do not receive the fringe benefits that a staff job provides.

Compensation varies according to the type of facility, where it is located, and the editor's experience. Independent post-production facilities often offer perks to highly regarded editors in lieu of larger salaries.

WORKING CONDITIONS

Videotape editing is very demanding, especially as you get better at your job. Clients pay a lot of money to have their shows edited, and they expect to get their money's worth from the people they hire. An editor must have the physical and mental stamina to be able to work extended shifts, even 20 to 30 hours at a stretch. He or she also must be able to handle pressure and be fast, efficient, and knowledgeable under any conditions.

An editor's family must understand that work sometimes comes first. As an editor advances and gains popularity, he or she might have to spend time away from home while working in other parts of the country or overseas. An editor must be dedicated to his or her career and decide whether he or she is willing to make the sacrifices required by an editing career.

SCREEN CREDITS AND AWARDS

Screen credits are the building blocks of your credentials in the industry, as they help people become familiar with your name and ability. Do not expect to see your name on the screen overnight. It might take years of editing before a producer or director will give you a screen credit. Sometimes you are given credit if you are working under a union contract whether the producer or director feels you should have it or not. Also do not expect a screen credit for every program you edit. I have more than 600 screen credits for film and tape editing, but I did not receive credit for about 200 programs that I edited.

Screen credits can lead to awards. The most coveted award in the television industry is the Emmy. Each year Emmys are given for outstanding achievement in videotape editing. An editor can win the award for his or her work on one show of a series or on a special.

Another way editors get recognition is by being involved in a highly rated program. The editor of such a show is often recognized by trade publications, which can be a tremendous boost to his or her career, as well as to his or her ego.

Perhaps the greatest feeling of satisfaction I have experienced is sitting at home watching a program I edited being broadcast to millions of people. It is exciting to see your name appear on the screen and suddenly realize "That's me!" Even though most viewers probably do not realize the importance of skillful editing, you can take pride in the knowledge that the program has your imprint on it.

The Client and the Post-Production Facility

13

This chapter discusses some basic points concerning client relations and standard capabilities and procedures found at many post-production facilities. Most of these points fall into that vague but crucial body of personal and administrative responsibilities that constitutes the nontechnical side of the videotape editor's job.

DEALING WITH THE CLIENT

Although the videotape editor's job is considered technical, as much as 80 percent of the time spent on a project consists of dealing with the client. If the client and the editor communicate well, the result will probably be a more successful project, in terms of both the quality of the editing and the money and time spent. Conversely, poor communication can severely undermine a project's efficiency and potential for success.

The editor must always be sensitive to the client's needs. The client usually is under tremendous pressure in terms of economics, scheduling, and the quality of the final product. Keeping the client's constraints in mind will help the editor understand his or her perspective. In addition, understanding the client's concerns will help the editor provide direction that will make the post-production process more efficient. When a project is run efficiently and the client's schedule and budgeting concerns are met, everybody wins. A satisfied client will most likely return again.

A videotape editor also must be patient. Verbally communicating an idea to another person, difficult in and of itself, will be even more difficult when the idea is a visual and possibly quite technical one. If you are not sure what the client wants or are unable to understand his or her questions, politely ask the client to restate the question in another way. Similarly, if the client does not understand something you say, rephrase the explanation or question. If both editor and client are patient, you will be able to overcome most problems in communication.

Unfortunately, the situation might not be so simple. Perhaps the schedule is very tight, you are exhausted, or the client is especially difficult. Although long hours are not very common except under unusual circumstances, they pose a real problem in dealing with clients. Not only does the editor's efficiency decrease rapidly, but every spoken word can become irritating. In these situations, it is even more important for you to remain as patient as possible. If you do not maintain a professional attitude, the immediate session, to say nothing of the project and possibly your job, will be jeopardized.

If an argument does occur, the fastest way to end it is to admit to being wrong, even if you are right. Sometimes it is difficult to convince the director or producer that you are right, especially if he or she might be embarrassed by being proved wrong. A basic rule of client relations is never to force your own ideas on a client—even if the product might ultimately be improved by your suggestion.

There are diplomatic ways to get a point across. If you believe that a client's edit will be a poor one, the best way to change his or her mind is to let the client see it his or her way first. If the client does not like the edit, offer an alternative. In most cases, the client will let you try it your way.

The client is paying the bills, and he or she is ultimately responsible for making the editing decisions. That does not mean you should refrain from making appropriate suggestions, but do so diplomatically. The client will usually appreciate your input, and after you make several helpful suggestions, the client will often solicit your advice. If your advice helps the show, the client might ask you to do another show or series.

An important but easily overlooked part of client relations is editing room posture. The attitudes and conduct of the editor and his or her staff are crucial in any facility. As a videotape editor, you must conduct yourself as a professional and have the technical and personal background to act confidently during a project. Depending on the client, it is sometimes helpful to break the tension in the editing room with a joke or two. Don't overdo it, though, and don't be a chatterbox. Most clients want to discuss only those things that directly concern the business at hand.

Finally, most editor-client problems occur in the first editing session, especially if the editor and client have never worked together. Whenever possible, it is a good idea to meet with the client before the first session to discuss the project. At this meeting, the editor and client will have the opportunity to establish the artistic and economic parameters of the project, and any potential problems can be ironed out before editing begins.

This meeting is also a good time to go over the proposed schedule to avoid any subsequent misunderstandings. For this reason, someone from the scheduling department should be present at the meeting. If it is a lengthy project, be sure the client is aware of possible overtime charges and union penalties. Schedules in the television industry frequently change for a variety of reasons, especially when the networks are involved. The client should understand how scheduling changes might affect the bill before making those changes.

CHOOSING A FACILITY

If the visual potential of videotape technology sometimes seems overwhelming to you as a prospective or working videotape editor, just think how a client, with a good deal less technical background, must feel. On a given project, a client will have to sift through all the different technical options, from the format to be used to the design of the titles, and then weigh these needs against the capabilities, service, rates, and quality offered by a number of different post-production facilities. Selecting a post-production facility can be a very difficult process.

Only the largest facilities will have all the capabilities described later in this chapter, and most of the others will offer basically the same type of equipment. Often a client must choose from among several facilities with similar technical capabilities, and the factors that distinguish one facility from another are service and personnel, rates, and the quality of the work. As a videotape editor, you obviously have control over the quality of the work you do. The other factors, although influenced by you and your performance, are often beyond your control.

A Facility's Nontechnical Capabilities

Service and Personnel. It is often easy to overlook the potential impact of personnel on a project. For example, an experienced editor might be able to do a particular job, but it might take considerably longer than the client expected or the facility might charge more because of the editor's experience. These extra charges might increase project costs to more than the client originally budgeted.

Aside from the personal character of each employee, the service offered by a post-production house seems to depend on three things: communication skills, general training, and specific training. It is important that the editor and other personnel in the facility possess the communication skills I discussed earlier in this chapter. General training refers to all the education a person receives prior to becoming a videotape editor. As I outlined in Chapter 12, this might include anything from a high school diploma to courses at a technical school or college. Specific training refers to the ability to operate the equipment available in a particular post-production facility.

One common problem found in videotape editing occurs when an unskilled or poorly trained editor tries to use an unfamiliar editing system. Before beginning a session, the editor should familiarize himself or herself with all the equipment in the editing room. Often an editor who is accustomed to working on one type of editing system will attempt to operate a unit made by a different manufacturer. Although different editing systems usually operate according to the same basic principles, the specific operation of systems can vary significantly. Efficient editing, especially when computer assisted, requires system-specific skills. A good editor should know the editing system so well that he or she can concentrate on creativity rather than mechanics.

The assistant editor's training also is important. He or she must be trained to anticipate the editor's needs, and this requires a thorough understanding of the editing system being used. Particularly important is the assistant editor's ability to coordinate the necessary videotape reels during an editing session. The longer it takes to set up and change reels, the more time is wasted. Remember, time is money, and the more money the client spends on staff inefficiencies, the less pleased he or she will be with the facility. A well-trained assistant editor helps increase the efficiency and makes the client feel secure in the knowledge that the editing will go smoothly.

Advance training of both editors and assistant editors is particularly important if free-lancers will be used. Many facilities use free-lance editors, audio mixers, and other technically skilled workers. To preserve its reputation, a facility should insist that free-lance editors and staff familiarize themselves with the relevant equipment before using it in an editing session.

Rates. Rates charged for videotape editing vary widely from facility to facility and from one part of the country to another. For example, in the Los Angeles area, probably the area with the most videotape editing activity, the rates for off-line time code–based linear editing systems vary from about $100 to $175 an hour. The most sophisticated nonlinear editing systems lease or rent for $2,000 to $2,500 a week. When broken down into an hourly rate, this averages about $50 to $60 an hour. Nonlinear editing systems are available only on a weekly or monthly basis because of the shipping, setup, and teardown time required.

On-line editing requires very sophisticated and expensive hardware, and the rates reflect this. An on-line session costs between $300 an hour for a simple edit to well above $1500 an hour for a complex session requiring digital effects.

A client should pay only for the items requested and for the amount of time the equipment is actually used. Each option is usually charged separately unless a package deal has been arranged with the facility. The hourly rate will depend on the amount of basic equipment and extras used during the session.

A client might be able to save money by guaranteeing a facility a certain number of hours of editing. Package deals are fairly common, and depending on the facility's schedule, the client might be able to save even more if the project is booked during a slower time of year.

Many small off-line and on-line houses can offer a producer a lower rate because of less overhead. Although this might sound tempting, the producer should make sure that the facility can provide the equipment and other resources required.

Preliminary Consultations and Technical Advice. Most post-production houses offer potential clients a free preliminary consultation or technical advice. This saves the client money and aggravation in the long run and promotes good feelings toward the facility. Even if the client chooses another facility for a specific project, he or she might return with a different project or at least refer other clients to the original house.

The client should choose a post-production facility before beginning production so that post-production preparations can be made well in advance and post-production requirements can be properly accounted for during production. Among the items that should be discussed during a preliminary consultation are how much tape the client plans to shoot, how much film the client plans to transfer to tape, and in what format the finished project will appear. Also important is the type of time code to be used (see Chapter 5) and whether the first edit of the program should be on-line or off-line.

One of the client's biggest concerns is making sure the project does not exceed the budget. A primary function of the post-production facility is helping the client meet the budget. A facility should make sure the client gets what he or she wants and not try to sell the client services he or she does not need. Through preliminary discussions, the editor can determine whether the client's expectations are realistic and can be met by the facility. The editor should stress advance planning and compatibility between the facility and the client.

A Facility's Technical Capabilities

Tape Formats. One of the first items to determine is whether the post-production facility can handle the client's planned tape format. One problem arises if the tape is shot in the half-inch Betacam or MII format. (Figures 13.1 and 13.2 are examples of recorders that use each of these formats.) Although these formats are popular for production, many tape houses are limited in their ability to use them effectively in editing. The reason for this is that both formats record the signal in a form called component video, which maintains picture quality through many more generations than the

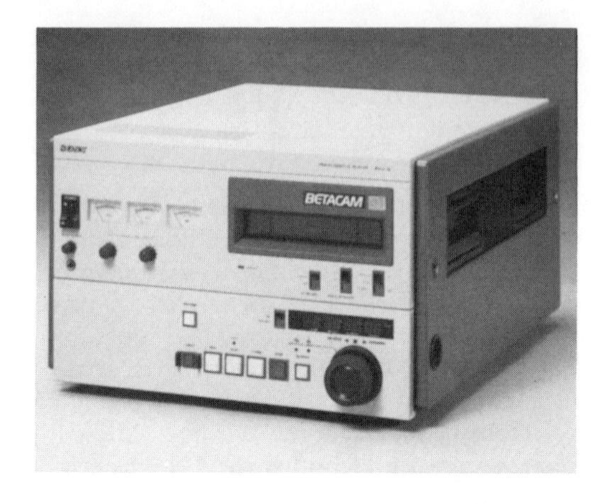

FIGURE 13.1
The Sony BVW-15 Betacam recorder.

standard NTSC composite video signal. To use these new formats in a cuts-only system, the output signal is converted from component video to NTSC composite. It then looks like any other video signal in a normal NTSC composite–based editing system, and it loses its higher video quality.

To take full advantage of Betacam or MII tapes, the video signal path must be able to handle component video. Assuming that the client wants fades, dissolves, wipes, and digital effects, this requires a video switcher that is able to pass the component video signal. The number of tape houses able to edit directly in the MII or Betacam format without converting to NTSC is increasing but still limited.

If a facility does have this capability, the client must know that he or she is currently limited to 20-minute cassettes for camera loads. Most people do not finish directly on MII or Betacam format but rather copy to one-inch type C videotape for broadcast. The use of the half-inch format as a mastering medium seems to be more practical now that half-inch formats are available in 30-, 60-, and 90-minute tape lengths, but only in studio models.

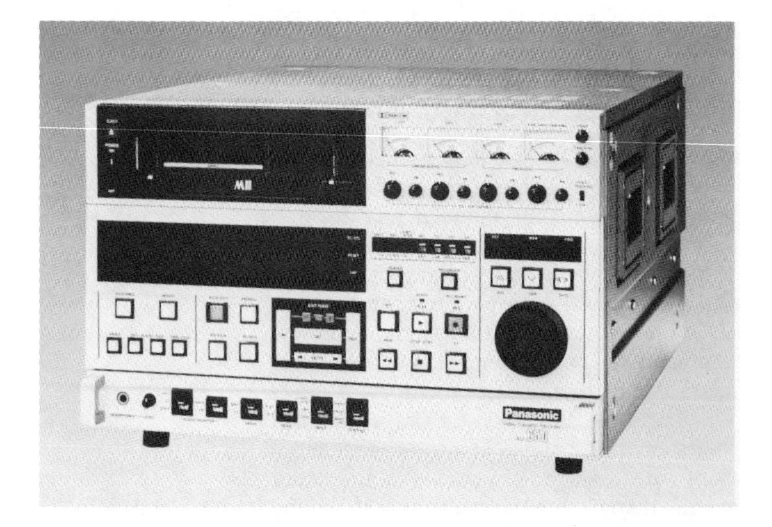

FIGURE 13.2
The Panasonic AU-650 MII recorder. (Courtesy Panasonic Broadcast Systems)

Film Transfers. Another key consideration is the integration of film and tape. Many facilities perform film-to-tape transfers or can assist a client in getting his or her film transferred before an editing session. If this service is required, the client will have to provide the facility with a specific kind of film that can be transferred to tape. Some of the things the client and facility need to know follow:

- Whether the facility needs a negative, positive, or interpositive
- What size film should be used (Super 8, 16mm, 35mm, or slides)
- Whether the film is silent or has sound
- What type of sound track is used (separate or composite, optical or magnetic, single track or multitrack)

Most houses that offer film-to-tape transfers also offer sonic cleaning, which removes lint, dust, and other types of surface dirt. Sonic cleaning will not eliminate scratches on the base side of the film or hide emulsion gouges. A few houses offer a wet-gate transfer in which a special liquid coats the film on both sides as it passes through the gate. This usually hides most of the base or celluloid scratches, but it will do very little to hide emulsion damage.

Graphics and Titles. Graphics, titles, and other forms of artwork are an integral part of most programs on videotape. There are three ways to add titles to a project—by putting artwork in front of a camera, by using an electronic character generator, or by using titles produced on film or tape prior to the start of editing.

If art cards or graphics are used, the post-production house will need to know the size of the cards to make sure it will be able to fit them on its camera stands. If the client provides titles already recorded on videotape, size and position are fixed. A digital effects device (at an additional charge) is required to change size at this point. The letter quality might be degraded by going through a digital effects device. It is usually more cost-effective for the client to bring in original artwork that can be photographed through a live black-and-white or color camera or to generate artwork through an electronic graphics generator.

With computer animation, a facility can create an incredible number of graphics variations and can even produce three-dimensional images. A number of excellent graphics devices are available, but they usually require a skilled graphic artist to realize their full potential.

If a product or actor must be photographed for a project, a small insert stage is required. Not all post-production facilities are equipped with insert stages and color cameras, and even if a facility does have this equipment, using them adds to the client's post-production costs.

Digital Optical Effects. A digital effects generator (Figure 13.3) manipulates the video image to create unusual special effects. If this type of special effect is needed, the client should plan it carefully on paper and discuss it with the post-production house's technical staff before going into the editing room to make sure what he or she wants is feasible.

Using a digital affects generator can add $150 or more an hour to the editing rate. Some generators also require the use of a special video switcher, which in turn might require additional videotape machines. All of this will add to the overall budget.

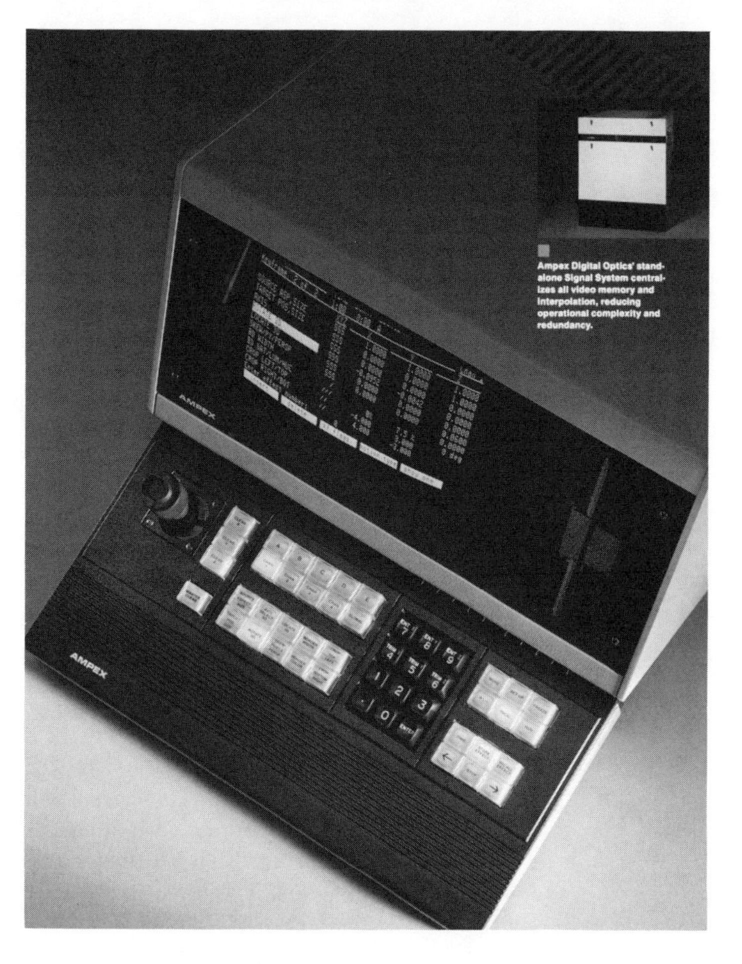

FIGURE 13.3
The Ampex ADO digital effects device.

CLIENT PREPARATION

Once the client has selected a post-production facility, the editor should make sure the client understands the importance of advance planning and preparation. This is an essential ingredient in the success of all projects, from creating the simplest off-line work print to producing the on-line master.

Time and financial considerations should provide enough incentive for the client to be properly organized. If the client slows down the editing process by making many changes during the conforming process or not being able to locate segments or whole rolls of tape, the charges will add up. The client must remember that the room has been booked for an editing session and that the clock does not stop while alternatives are discussed or misplaced tapes are located.

I have already discussed the importance of maintaining accurate records (see Chapter 4), and this information should be passed along to the client if necessary. In addition, the client should make as many editing decisions as possible before the editing session begins. This process is referred to as a paper edit because decisions are written down on paper prior to the off-line or on-line session.

Another means of organizing material for post-production is using a storyboard, which is a series of individual drawings or photographs that depict the continuity of the program. Storyboards are particularly useful when the post-production process is complex.

The best way for a client to make editing decisions and generally prepare for post-production is to view videotape work prints with the time code numbers printed in the picture (window dubs). The post-production facility can usually provide these. The client can view the window dubs on a low-cost half-inch or three-quarter-inch VCR that provides still-frame and frame-by-frame access of the picture. He or she will use the window dubs to record time code information concerning specific edits. The client's editing job will go faster if he or she brings in a rough list of time code numbers in editorial continuity or, even better, a frame-accurate edit list that is ready to be off-lined or auto-assembled.

Planning pays off not only in terms of time and money, but also in creating a better, more coherent final tape. With the editor's guidance and help in planning, the client's project has a much better chance of being successful. If, despite the editor's advice and warnings, the client is unprepared, he or she will not be able to blame the editor or post-production facility if the project's schedule or budget is threatened.

THE OFF-LINE PROCESS

Three basic types of off-line editing are available. The first is the client's use of one or more low-cost VCRs to view his or her material. Although this equipment does not edit material together, it does allow a producer or director to select material and make preliminary choices at the lowest possible rate.

The second type of system is a relatively simple cuts-only, two-VCR setup. This system will generate an edited work print on a ¾-inch cassette or in Beta or VHS half-inch format. Some basic cuts-only editing systems make edits by using control track pulses on the tape to identify frames. As noted in Chapter 5, this type of system is not frame accurate, but a time code option can be added to the system to read time code data recorded on the work tape. This will result in an edit list that is essentially frame accurate. Most new systems offer the option of connecting a paper punch tape or a less expensive hard-copy printer to store edit decisions. The punch tape provides an edit list in a format that is compatible with most on-line systems.

The third type of off-line editing system is an expanded version of the second. Three or more VCRs are included in the basic system, which also includes a video and audio switcher to create optical effects such as wipes, fades, and dissolves. Also included is equipment to store edits on either paper tape or magnetic floppy disks. The system can be connected to a low-cost printer to provide a hard-copy readout of all editing decisions. This system uses time code to guarantee frame accuracy.

Computer software programs that operate on a personal computer are now available. This allows the user to generate industry standard EDLs on 5¼-inch or 3½-inch floppy disks, which may be converted by most post-production facilities to the standard 8-inch format or, in some cases, use the client's small format disk without conversion.

Once off-line editing is completed, the next step is having the edit list cleaned (see Chapter 8), which includes removing unwanted edits and arranging the remaining edits in an efficient order for auto-assembly. Some newer off-line systems offer list cleaning as an option, and most post-production facilities have software to clean off-line edited lists. The charge for list cleaning is between $50 and $75 per hour, and most lists can be cleaned in less than an hour.

Up to this point, the production is considered in the work print stage of editing in which relatively low cost equipment is used. Editing rates will vary according to whether time code is added, whether the facility supplies its own editor, and how many VCRs and other options are requested. Time code editing systems are more complex to operate than control track editing systems and usually require an editor who is skilled in the operation of the particular system.

THE ON-LINE PROCESS

On-line editing refers to the process of creating a videotape of master quality. The term *on-line* does not automatically mean editing a one-inch videotape. Half-inch Betacam and MII tapes, as well as three-quarter-inch tapes, are recorded in the field and are either bumped up to one-inch or are used as play sources when the master is a one-inch videotape.

On-line facilities offer digital effects generators, title generators, cameras, and audiotape recorders that might be needed during an editing session. All these items should be booked before the session.

Many smaller post-production houses use their off-line editing systems not only for building cassette work prints but also for a form of editing known as mastering. A cassette master is any type of edited material that does not have time code numbers in the picture and can be used for viewing purposes. Mastering is another name for on-line editing that uses lower cost (and sometimes lower quality) tape formats. In some cases, the cassette master might not be broadcast quality, but if an actor or actress needs a demo tape to show to agents, a director wants to put together a series of commercials, or an editor wants to build a sample tape of his or her best efforts, mastering is the best method to use.

POTENTIAL PROBLEMS

Problems of one sort or another always arise in post-production. The most the editor can hope to do is minimize the number of problems through planning and communication and, when problems do occur, deal with them in the best way possible. The following problems are the most common ones faced by videotape editors.

Client Errors

Whether a client has carefully planned a project or not, he or she will make errors in organization and decision making. Some errors can be solved easily, but many can slow up or even stop an editing session. If the client has been doing business with the facility for a long time, it might be good business not to charge him or her for the downtime resulting from poor organization. Many times a good word from this client will generate more than enough business from other clients to offset any inconvenience to the facility.

Client-Supplied Stock

Some post-production houses allow clients to supply their own tape stock for transfers or editing. Although clients can save money by buying stock in large quantities and bringing it with them to an editing session, an entire session might have to be scrubbed if the client happens to get a defective piece of tape.

There are three reasons why post-production houses charge more for their tape stock. First, it is evaluated. Second, it is preblacked when required for editing. Third, if facility-supplied stock is defective, the facility absorbs the cost of redoing the project. If client-supplied stock is defective, the client is responsible for costs incurred in redoing the project.

Schedule Changes

Schedules can change quickly in the television industry. A network might decide to air a show earlier than planned to help achieve better ratings. Or a show without an air date might suddenly be scheduled without leaving enough time for its completion. Both types of problem force the client and the editor to rearrange their schedules and those of all of the subcontractors, and they usually result in long, exhausting work sessions.

What can be done? If the editor and the facility cannot rearrange their schedules, the client will have to go elsewhere. Often, however, another client will agree to change his or her schedule to allow the facility to accommodate the first client. Although it is good practice to try to accommodate sudden schedule changes, the client must solve this problem on a case-by-case basis.

Downtime

Downtime as described in this chapter is the non–revenue-producing part of a scheduled editing session. Of all the problems that can cause tension in the editing room, downtime is one of the worst. Downtime is usually caused by the malfunction of equipment critical to the system's operation. Minor problems such as the failure of one VTR might not stop the editing session, but they can slow things down by forcing the editor or assistant to change reels more often. This is less of a problem during off-line editing, but in on-line editing, when broadcast tapes must be set up and balanced each time a reel is changed, it can slow down the process considerably.

The first thing most clients complain about is how much of their time is being wasted, even though they are not being charged during downtime. If the repair time runs into hours, it upsets not only this client's schedule but also other clients waiting to get in. If the facility cannot offer the client extra time right away, the client might decide to abort the project and go elsewhere.

As previously noted, downtime also can be caused by the client. A good example is when the client fails to have all the videotapes or other material on hand. If the client must stop the session to get a tape or other material vital to the job, it is up to the facility to decide whether to charge the client for the downtime.

Another type of downtime occurs when the VTRs or other equipment in the editing room are used for preblacking tapes, duplicating tapes, or performing some other nonediting operation. This time also can be used for routine maintenance. All these functions generally occur during slow times, usually after 6 p.m.

An editor's carelessness can result in downtime. If an editor leaves without restoring the equipment to its normal condition, another editor might find that the editing room is down simply due to a patch cord in the wrong place or a piece of equipment that has been moved out of calibration to perform a special function. Most post-production facilities have the maintenance department normalize every piece of equipment in the editing and equipment rooms each morning and also verify that it is in good working order. This type of preventive maintenance is one way to minimize downtime and keep the facility profitable.

Equipment Breakdowns

When a technical problem arises during editing, the editor should find a maintenance engineer as quickly as possible and explain what the problem appears to be. Then it is up to the engineer to solve the problem. The only thing the editor can do is reassure the client that everything is being done to resume the session as quickly as possible.

If the problem appears to be minor, the client can probably take a short break for coffee or make phone calls while waiting for it to be fixed. If, however, the problem appears to be extensive, the maintenance engineer might be able to give the editor and the client an idea of just how long he expects it will take to solve the problem.

When the equipment is down, tension can build, and the editor's diplomatic skills will be put to the test. The editor should get the client out of the editing room until the problem is solved and try to reassure him or her that the engineer is doing everything possible to solve the problem.

Faulty Time Code

An entire day can be spent correcting poor time code. If at all possible, have the client bring in the tapes before the start of the editing session so the staff can spot-check them. Normally, time code problems occur when tapes are made in the field or when a small production facility not used to recording time code is asked to supply it. Nearly all facilities using time code for editing, however, are aware of the requirements for time code recording and will be able to supply good quality time code.

Computer Errors

If the editor is working with a computerized editing system in which an edit list is being generated, an edit error might cause a problem if it accumulates over a number of edits. If the problem can be corrected after the client leaves, it need not be mentioned, but if it affects the continuation of editing, the client should be told immediately.

CONCLUSION

As you can see, many different factors, both technical and nontechnical, affect the post-production process. The editor and the post-production facility must be prepared to play a number of different roles when dealing with various clients.

New Technology and the Future

14

Since the beginning of the 1980s, the television industry has experienced dramatic technological changes. In this chapter, I discuss this new technology and examine the immediate future of videotape editing.

LOCATING MATERIAL

The most time-consuming part of videotape editing is finding and locating material. Many three-quarter-inch VCRs used in editing today take more than 12 minutes to go from one end of a 60-minute cassette to the other. (See Figure 14.1.) Even the most sophisticated of these machines take more than four minutes to perform this task.

Currently, one-inch broadcast VTRs search at speeds of 50 times play speed or greater, but the majority of editing time is still devoted to threading and unthreading VTRs or searching back and forth until the desired material is located. Even at 50 times play speed, it takes well over a minute to go from one end of a 60-minute tape to the other. This more than two-minute round trip multiplied by dozens of edits equals a lot of expensive editing time.

Indeed, locating material requires the most time during the editing process, with decision making coming in second. Making the actual edit takes the least amount of time.

How can editors reduce search time? First, adequate notes must be taken during production to allow the editor to go to some specific scene or take based on time code information or slate identification. Those notes could then be used in conjuction with a mass-storage medium such as an optical laser videodisk interfaced with the appropriate edit controller. Access to a particular edit point would then be reduced to a fraction of a second instead of well over a minute. Decision making would be greatly improved, and I suspect that the creative quality of the product also would be improved because the editor could maintain a sense of continuity by not having to wait for tape machines to find the material.

THE CMX-600

Believe it or not, the CMX Corporation introduced a system that would do all this in 1971. Although crude by today's standards, the CMX-600 (Figure 14.2) could locate any frame of information within its 27 minutes of magnetic

FIGURE 14.1
A Sony BVU-800 three-quarter-inch VCR used in off-line editing. (Courtesy Sony Corporation.)

FIGURE 14.2
The CMX-600 editing console, a magnetic videodisk-based system. (Courtesy CMX Corporation.)

videodisk storage banks in one seventieth of a second. That's about twice as fast as the blink of the human eye. For all practical purposes, this system provided instantaneous retrieval of picture and sound information.

Why is the CMX-600 no longer around? There are a number of reasons. First, its initial cost was about $350,000. Second, because of its high maintenance costs, only a few facilities could afford to keep it running. Third, the CMX-600 produced only mediocre pictures in black and white. Finally, it had a storage capacity of about 27 minutes, limiting the amount of raw material the editor had to work with.

THE CMX-6000

In 1986, the CMX Corporation introduced a new editing system based partially on the CMX-600. The CMX-6000 (Figure 14.3) uses random-access color laser disks that allow the editor to see and study frames individually and to cut action and dialogue with precision. Rearranging scenes and cuts is very easy. In addition, the editor can jog the audio frame by frame and listen to each frame of sound discretely. One interesting feature of the CMX-6000 is its ability to cut picture and sound independently.

The basic 6000 system, unlike the earlier CMX-600, has a capacity of 30 minutes of picture and sound storage, but the system can be expanded to contain 2 hours of picture and 4 hours of sound material. The price is about the same as that of a conventional eight-plate flatbed film editing system, about $50,000, which is a far cry from the $350,000 price tag of the earlier model. Even the laser disk players cost a fraction of what the old disk drives did.

A keyboard is included with the system to allow the editor to log in scene and take information. This same information can be loaded into the CMX-6000 from an IBM PC or PC-compatible computer.

This and other systems like it replace the labor-intensive process of film editing with state-of-the-art videotape editing hardware and software that perform editing functions at lightning speeds. I am convinced that any film editor who learns how to use this new technology will never want to go back to commonly used film editing techniques.

FIGURE 14.3
The CMX-6000 laser videodisk editing system. (Courtesy CMX Corporation)

DISK STORAGE AND ACCESS

Because of equipment such as laser disk players, we have the ability to randomly access high-quality color picture and sound information instantly. Further, high-density laser disks will no doubt extend the storage capacity of editing systems to one hour, offering the added advantage of recording on both sides of the disk and thus giving editors two hours of raw material per disk.

One feature that is already available in the Ediflex, a videodisk-based editor, is light pen control. With this system, editors use common English words as editing commands (for instance, the user can call up material by scene or take number). Thus, although this type of system is time code based, the time code is not visible to the user. Another new feature is the ability to edit at 24 frames per second, allowing simple conversion back to film edge numbers for negative cutting.

What are some of the features that would give the new generation of video editors even more flexibility? One is a videodisk player that has a minimum capacity of 30 minutes per side, for a two-side total of 60 minutes. Other desirable features are the ability to scan the video and audio at any rate selected by the user, the ability to freeze a clean, usable field or frame indefinitely, and the ability to make edits from disk to disk or add material to the same disk. Of course, future editing systems must offer instant random access of any material recorded on any disk within the system, and they must be reliable and easy to use. The industry doesn't need another piece of hardware that requires a staff of engineers to maintain it.

Videodisks are also called laser disks. Laser disks currently being used for editing contain up to 54,000 tracks, or frames, per side, which translates to a maximum capacity of about 30 minutes per side. Each track contains one picture frame. In still-frame mode, the laser reproducer scans one track at a time, displaying one picture frame indefinitely.

Figure 14.4 shows a laser disk recorder. This device records up to 30 minutes of color video, sound, and time code in real time.

Some 60-minute laser disks are currently available, but these record two picture frames per track, which can pose a serious problem. If an editor wishes to access only one frame on this type of disk, the player will attempt to play both frames. Advances are being made, however, and laser players that use high-density disks with 108,000 tracks per side will be able to access each frame independently.

Another challenge for the future is overcoming three additional limitations of current videodisk units. One is the shortage of facilities capable of transferring material to laser disks for editing purposes. The second is the inability to get a videodisk made in a short time. You cannot use a videodisk as a source in post-production if the turnaround time is not within the show's schedule. The third problem is the cost of laser disks used in editing.

One possible answer to these concerns is Sony's development of a reusable (or erasable) laser disk. A reusable disk would virtually eliminate the need to have laser disks made by an outside contractor. Sony and several other manufacturers also are in the process of developing low-cost, practical laser disk recorders that can duplicate disks in much the same way tapes are copied today. Cost estimates range from about $35,000 to $75,000 per unit, and such devices are not expected to be available for post-production until late 1990.

FIGURE 14.4
An optical disk recorder used to make videodisks for systems such as the CMX-6000 and the Spectra System. (Courtesy Optical Disk Corporation)

ADVANCED EDITING SYSTEMS

Spectra System

The Spectra System from Spectra Image (Figure 14.5) is somewhat different from most laser editing systems in that Spectra has designed its own laser disk player with two heads. Each editing system is able to use up to five of these dual-head players, in effect providing a capacity equivalent to ten disks, with up to two and a half hours of material from which the editor can select.

The model D220 high-speed, dual-head laser disk player (Figure 14.6) is able to search through up to 30 minutes of material and cue to any point on a laser disk in less than 1 second. Each player has two playback heads mounted 180 degrees apart that act as independent units. This allows the system to independently access the same surface of a single-sided disk simultaneously and independently. Since both heads are under computer control, while one is playing, the other is searching or may already be cued for its next edit. One cost-effective factor in using dual-head laser players

FIGURE 14.5
The Spectra System, which uses dual-head laser videodisk players. (Courtesy Spectra Image)

is that the number of laser disks required for editing is cut by 50 percent.

Because of the use of dual heads, dissolves, wipes, and fades can be made from material on the same disk. Unlike other laser disk–based editing systems, which require multiple copies (up to three disks) to achieve random accessibility, the Spectra System requires only one disk for each 30 minutes of material.

As with other types of random-access video editing systems, the ability to cut and paste scenes or sequences and other methods of rearranging material are all available to the editor. Although this is a time code–based system, the time code is transparent to the user. The Spectra System also includes proprietary audio hardware and software that logs level changes and mixes so that they can be accurately repeated in the on-line conforming process.

FIGURE 14.6
The dual-head laser videodisk players used in the Spectra editing system. (Courtesy Spectra Image)

Montage

Another approach to random-access editing is illustrated by the Montage Picture Processor (Figure 14.7), a device claiming to be more efficient than the laser disk. Unlike laser disk–based editing systems, the Montage uses up to 17 Beta Hi-Fi VCRs. Each VCR stores up to five hours of picture and sound information, and all 17 VCRs in the system store the same information. When edits are made and played, each VCR searches for the needed shot and waits for a cue. At the appropriate time, one or more VCRs go into play and show the edit.

Thus, when a series of edits has been preprogrammed by the editor, all the VCRs cue up automatically and after a few seconds go into play. Because the same material is recorded on 17 VCRs, each machine, controlled by the system's software, knows where to go. The pictures and sound are switched in real time, and when one VCR finishes playing its edit, it quickly recues to a new edit point assigned to it by the computer. The VCRs are constantly shuttling back and forth during a playback sequence of edits, allowing the material to be viewed in real time, edit after edit, without stopping.

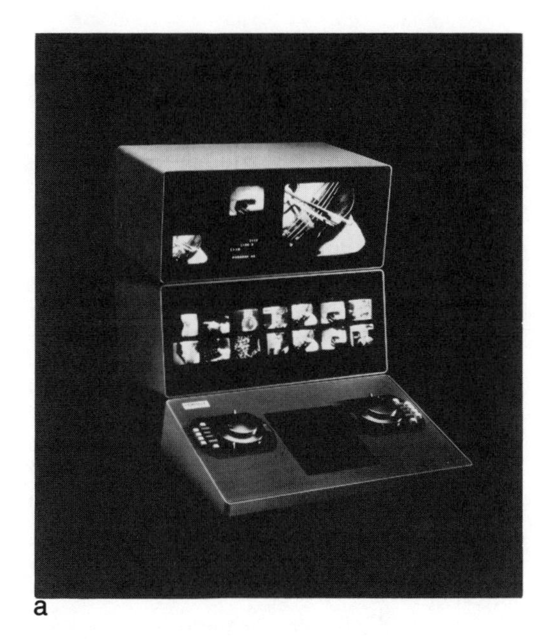
a

b

FIGURE 14.7
The Montage Picture Processor (a), which accesses up to 17 Beta format VCRs, and a module (b) of 7 VCRs. (Courtesy Montage Corporation)

Under average circumstances, all the edits can be viewed on the fly. On occasion, when a VCR cannot get to the edit point in time, the system stops, recues, then continues to display programmed edits. The reason for the apparent random accessibility of the Montage is that it does not record a videotape work copy of each edit. Instead, the edit list time code data is accurately stored in memory, and when the editor is playing a series of edits, both video and audio edits are switched in real time. These edits are generated from the data stored in the computer's memory and thus can be replayed over and over again without recording them on videotape. The internal edit list is constantly updated so that when a sequence is approved, the list is instantly ready for automatic assembly.

Ediflex

According to the manufacturer, Cinedco, the Ediflex (Figure 14.8) is a dialogue-based editing system. That is, the editor is able to number lines of dialogue in the script and reference them to a log in the Ediflex. The editor need only call up those lines of dialogue he or she wishes to use by number, and the system will assemble them in the correct order.

Like other nonlinear systems, the Ediflex was designed with the film editor in mind. Its data base is script-oriented and stores every line of dialogue and every camera angle in its memory. The Ediflex system is comprised of a disk-driven 64K computer, data display screen, and three-quarter-inch JVC VCR, which loads material into eight JVC VHS machines. The edit list is stored in SMPTE time code format on an eight-inch floppy disk. The Ediflex will output code numbered or key numbered edit lists for conformation back to film.

Availability of Systems

At this time, the Spectra System and the Ediflex are not available for sale but can be rented or leased for short- or long-term use. The Montage can be rented or purchased. The selling price is $125,000 and up. Although none of the systems is cheap, using electronic editing systems can cut film post-production time by more than 50 percent, which translates to less expense in the long run.

FIGURE 14.8
The Ediflex-12 multi-camera video editing system for fast-action programs. (Courtesy Cinedco.)

WHERE DO WE GO FROM HERE?

The enormous technological advances of the 1980s are leading the television industry into new and exciting territory. One such advance is the use of high-definition television to produce commercials, music videos, and feature films. The high-definition process results in more than four times the picture resolution available from regular television. As noted in earlier chapters, the NTSC television standard is a 525 line video picture frame being scanned at 30 frames per second.

High definition video has an 1125 line video frame, but is also scanned at 30 frames per second. Since the frame rate of high definition television (HDTV) is the same as NTSC television, existing editing hardware may be utilized with very little modification. In essence, HDTV is a high resolution version of NTSC television. Being able to use conventional video editing equipment to edit tapes generated by the high-definition process is a definite plus for today's editors. Feature film producers can use high-definition television to take advantage of the complex optical effects available with electronic editing and still have picture quality equal to that of 35mm film.

High definition video tape, when transferred to 35mm color motion picture film, in some instances is indistinguishable from the same scene photographed with a 35mm motion picture camera.

A Canadian mini-series, "Chasing Rainbows," was produced on HDTV videotape and later the finished product was transferred to 35 mm color film. The transfer from tape to film produced a color print of quality that was hard to distinguish from direct optical photography.

Within the next five years, mass storage devices such as bubble memory might replace videotape in many applications. Vertical magnetic recording is a high-density recording process capable of storing up to 20 times the amount of information previously possible on laser disks, floppy disks, and hard disks.

Smaller disks also are the wave of the future. We will soon be able to use the familiar 5-inch compact laser audiodisk for video as well as for audio. Eventually, these smaller disks will provide a much less expensive source of laser-read video material for editing systems.

Another area of the video industry currently undergoing rapid development is the production of digital video effects (DVE) generators. Two such devices are Quantel's Mirage and Ampex's ADO (Ampex Digital Optics). Although both devices are similar in operation, they can do different things and can be connected together to create some very unusual optical effects.

Whatever the type of technology, an important part of videotape editing in the future is sure to be more user friendly systems. Editing is an aesthetic activity, and editors should not have to go through a series of complicated steps to make a simple edit. At present, it is difficult to communicate with some of the popular editing systems, but newer, more accessible systems using sophisticated software have greatly reduced the intimidation factor.

In the future, manufacturers must thoroughly investigate the needs of their potential customers and try to make the editing process more efficient. The clue is to build systems that allow the editor to think more about the aesthetics of editing and less about the mechanics of performing the edit. When compared to today's technology, these new tools will make the task of post-production easier on the editor and certainly on the budget.

Glossary

Action Match The smooth movement from one edit to the next.

Address Track The auxiliary audio channel on videotape most often used to record time code information. Also called the *cue channel.*

Alphanumeric Of or relating to the 26 letters of the alphabet (uppercase and lowercase), the numbers 0 through 9, and special characters such as #, $, %, *, (, and).

ASCII American Standard Code for Information Interchange. Denotes a list of encoded letters, numbers, characters, and symbols similar to those on a standard typewriter keyboard. *See* Alphanumeric.

Assemble A mode An automatic sequential assembly of a series of edits.

Assemble B mode An assembly of a series of edits performed nonsequentially in checkerboard fashion from one reel of a source videotape at a time.

Audio-follow-video A special function found on some video switchers (usually those used in production applications) that allows the tracking of sound associated with video signals, such as those from a camera, each time a switch or optical effect from one video source to another is required.

Audio mixing *See* Sweetening.

Audiotape recorder (ATR) An eletromechanical device for recording or reproducing sound. This includes any format, including cassette, quarter-inch, and larger formats.

Auto-assembly A generic term used to describe the computer assisted automatic assembly of a series of edits from information stored in the computer's memory.

Automatic dialogue replacement (ADR) A function of the sound editing and mixing process used to replace unusable dialogue recorded during production with dialogue rerecorded under carefully controlled studio conditions.

AUX or AX (Auxiliary) An input source in a videotape editing system used to connect television cameras, audiotape recorders, film chains, or other devices not under time code control.

Billboard A form of sponsor identification that can be used at the beginning or end of a program or going into or out of a commercial break. A billboard is usually very short, lasting from three to ten seconds.

Binary Having two states, such as on and off or positive and negative. Binary codes are composed of 1s and 0s. SMPTE time code is generated in a binary form. Computers convert binary data into letters and numbers.

Bit The smallest unit of electronic information. Computer data is made up of many bits of information. Eight bits make up one byte, the equivalent of one letter or number.

Black compression The loss of detail in the dark areas of the picture. For example, black compression might occur in a shot of someone's clothing in which the pockets, lapels, and fabric design are indistinguishable.

Black level A reference point in the video signal (picture) that is nominally 7.5 percent above zero video, or the blackest point in the visible picture.

Bloop A term used to describe a missing piece of sound (*see* Dropout). In most cases, it is used to eliminate unwanted sounds.

Bumper A short logo or graphic used to remind the viewer of the program name when going into or coming out of a commercial break. It can be silent or contain theme sound.

Bus An electronic connection between devices or part of a single device that transmits and mixes input or output signals from a number of sources.

Byte The smallest unit of information that can be addressed by a computer. A byte is made up of eight bits and is the equivalent of a single letter or number.

Character generator (graphic type) A device allowing the user to create letters, numbers, or graphic figures and combine them with a picture. This type of character generator is used to generate titles and roll titles, crawls, or page-by-page text.

Character generator (time code type) This device is used in videotape post-production to generate visible time code numbers that are added to video information to aid in establishing edit points. The time code character generator is used primarily in the off-line editing process. A picture with time code numbers inserted in it is known as a window dub.

Chroma The amount of color information in a television picture; more simply, the intensity of the color. Low chroma means that the color picture looks pale or washed out. High chroma means that the color is too intense and has a tendency to bleed into surrounding areas, contaminating other colors.

Clean edit list An edit list that has been processed manually or with a special computer program designed to remove unwanted edits and reorder the list in an efficient manner for automatic assembly.

Color bars An electronically generated set of video reference signals consisting of vertical blocks of the following colors: red, green, blue, magenta, yellow, cyan, gray, white, and black. Videotape machines, cameras, telecine chains, and monitors all use color bars as an absolute reference for proper picture setup.

Color burst A color reference signal that is part of the composite video signal when color information is being displayed. The burst signal has an amplitude of 40 units and can be viewed on either a waveform monitor or a vectorscope.

Comment *See* Note.

Component video The separation of the luminance, or video, and the color, or chroma, parts of the television signal. These two signals are recorded separately on the videotape. This separation helps maintain better picture quality through more generations. Component video signals require special cameras, recorders, and video switchers.

Composite Of or relating to the picture and sound track combined on the same piece of film or tape. Most videotape is edited in the composite form. *Composite* also refers to the combination of various picture or sound track elements into a single picture or sound track.

Composite video The total video signal, consisting of 40 units of synchronizing and blanking signals and 100 units (equivalent to 100 percent) of video information, including color information. When properly adjusted, the balance between the blanking signals and the video information should add up to 140 units.

Conforming The matching or copying of information from one source to another. Film negative cutting is one form of conforming. The term is most commonly used in videotape editing to describe the copying of material from one video reel to another according to a list of editing decisions based on frame-accurate time code information.

Contrast The difference in intensity between the light and dark areas of the picture. A picture with high contrast has extremes of intense white and solid black, while a picture with low contrast has no extremes, just varying levels of gray. A picture with normal contrast contains a good rendition of blacks and whites and includes all tones of gray.

Control track Electronic pulses recorded on videotape to maintain the constant playback speed of the tape. Each video frame has one pulse. Missing control track signals can cause the videotape to break up or mistrack. This is analogous to torn sprocket holes on film. Control track pulses are used in simple editing systems as reference points to make edits.

Cue channel See *Address track.*

Cutaway A shot used to maintain the continuity between two scenes. Also called an *insert shot.*

Cuts only A low-cost editing system designed only to make cuts. No dissolves, wipes, fades, or other optical effects can be made on a cuts-only system.

Dailies A motion picture term also used in television to refer to film production material photographed on one day, developed and printed immediately, and viewed the next day. Dailies on videotape are electronically recorded and therefore require no processing, so they can be viewed as soon as the tape is rewound.

Digital effects generator A device capable of manipulating a video frame in still-frame mode or in motion in just about any way. With a digital effects generator, a single frame can be enlarged, reduced to infinity, pushed off the screen, flipped, twisted, or split apart and put back together in real time.

Digital signals Electronic signals defined by the presence or absence of a specific voltage. The two possible voltage states constitute a positive (on) or a zero (off) signal. Sequences of digital signals embody information that can be stored and transmitted electronically. Computers and related technology use digital signals.

Dissolve A slow transition from one scene to another created by superimposing the incoming scene over the outgoing scene, gradually decreasing the strength of the former and increasing the latter until the incoming scene is in its full intensity and the outgoing scene has disappeared.

Distribution amplifier (DA) A device used to maintain constant audio or video signal levels so that they can be simultaneously distributed to several other devices.

Double system An editing or projection system composed of two separate units, one for sound and the other for picture. Motion picture film is most conveniently edited on double system equipment. Videotape, however, is a single system device, with both picture and sound recorded in synchronization on the same piece of tape.

Downstream keyer A device installed in a video switcher to control the video signal just before the final output stage of the switcher. It allows the editor to add titles or other graphics without tying up one of the switcher's other video effects buses.

Downtime A television industry term describing the suspension of an editing session due to a technical or other problem. Downtime also can be defined as nonproductive time, such as time when technical equipment is operational but idle.

Drop frame time code A type of time code developed by the SMPTE to make the length of the time code on the videotape agree with clock time. Chapter 5 contains a detailed discussion of this subject. *See also* Non–drop frame time code.

Dropout A momentary interruption of sound or a flash or other discontinuity in the picture. Dropouts in video are seen as tiny black or white horizontal flashes.

Dropout compensator A device designed to replace a missing line or part of a line of video with adjacent video information so the dropout will not be detectable.

Dubbing *See* Sweetening.

Duration The length of time required to complete an effect such as a wipe, fade, or dissolve. Also, the length of an edit, starting at the beginning time code number and stopping with the end time code number.

Edit A defined segment of picture and/or sound that is assembled with other similar segments to build program continuity on videotape or film. The term also refers to the act of selecting and assembling these segments.

Edit decision list (EDL) A computerized list of time code numbers, notes, and related data that specifies a sequence of edits. A computerized editing system can use such a list to assemble edits on a separate videotape.

Edited master The product resulting from an electronic editing session. An edited master is usually generated from original source material. The edited product is also called a *second-generation master* or *electronically edited (E-E) master*.

Editing, electronic A method by which television pictures and sound recorded on one videotape are transferred to another by electronic means. This new edited copy is regarded as a second-generation tape.

Editor A person who builds a sequence of scenes that flow together creatively. This person is also in command of the equipment and the editing staff and relates to all the people involved in the project. The term *editor* also is used to refer to a computerized editing system.

Effect An effect is a transition between edits that involves more than a straight cut. Examples are a dissolve, wipe, fade, or special digital effect. The edit starts at the transition point and continues until the effect has been completed. The length of time it takes to complete an effect is called the duration.

Electronically edited master (E-E master) The edited videotape master, usually at least second generation, that is used to make copies for duplication or broadcast.

E-MEM A trademark of the Grass Valley Group. A memory storage system on a video switcher that allows complicated switcher effects and transitions to be stored, recalled, modified, and repeated with precision. The term is often used generically to refer to all such memory storage systems.

Error message A message generated by computer software that notifies the operator of a failure to execute the instructions given to the computer.

Fade An audio or video transition that goes from full intensity to silence or black or vice versa to begin or end a sequence.

Field One half of a television picture frame. In the NTSC television system, each television frame is made up of two distinct fields. Each field is made up of 262½ scan lines, and together they provide the full 525 scan lines used in the NTSC television system. There are 60 television fields per second (two per frame), or 30 television frames per second.

Film chain A group of up to three film slide projectors or special motion picture projectors connected by a mirror arrangement, allowing the switchable selection of any one projection device into a special television camera unit for recording on videotape or live broadcast. The camera can be one designed for telecine use or a conventional television camera suitably positioned to pick up the image sent by a projector. Some film chains even have a Super 8mm television projector included, but they are not as common as 16mm and 35mm units.

First cut *See* Rough cut.

Flash frame An extra field or frame of video usually seen at an edit point. A flash frame can result from a system that is not frame accurate. The only way to correct this type of problem is to remake the edit subtracting a frame from either side of the edit point. Using a preview mode will help the editor determine which side of the edit contains the flash frame.

Flying spot scanner A film-to-video transfer device that scans the film frame in a continuous motion rather than using the conventional intermittent pull-down method.

Frame A television video frame. Each frame is made up of two fields, one odd and one even. *See* Field.

Frame store A digital device that stores in memory a single field or frame with good quality for an indefinite period of time. *See* Freeze frame.

Freeze frame A still picture of a single television field or frame. Also called a *still frame* or a *hold frame*.

Head, audio A nonrotating device used on a VTR or VCR that is usually almost identical to the heads used on other audiotape recorders. The audio head may be a combination record and play head. A separate erase head is usually provided to erase the tape before recording.

Head, video The part of a VTR or VCR that records or plays back the picture portion of a videotape signal. A video head is an electromechanical device, usually part of a VCR or VTR, designed to record and play back video signals. It is mounted on a scanner, which holds the head firmly in place while scanning the tape. Also called a *video scanner*.

Helical scan A VTR or VCR recording format that wraps the tape around the video scanner in a helix pattern. The video scanner rotates at a slant to the tape and thus records the video information at an angle. Also known as *slant track*.

Hertz (Hz) A measurement of frequency defined as cycles per second.

Hold frame *See* Freeze frame.

H-shift A slight horizontal shift of the picture to the left or right of the screen at an edit point. An H-shift occurs when the editor is creating a match cut and is caused by either a color framing error or an improperly adjusted time base corrector. The shift might not be very large but is noticeable even to the untrained eye.

Hue *See* Phase, color.

In point The beginning of an edit defined by an eight-digit time code number. The in point is where the start of one scene is spliced to the end of the previous edit. Also called *in time*.

Input/output (I/O) device A device used to connect several electronic devices together. For example, an I/O device might be used to connect a printer to a computer so that useful data can be printed.

Insert editing In essence, insert editing replaces an existing video signal by recording another video signal over it on a frame-for-frame basis. This is the method used most frequently in video editing.

Isolated camera A single television camera dedicated to a separate videotape recorder and run continuously throughout the shooting of a scene. It is only one of several cameras used during shooting and is not usually considered the primary camera angle. It is used to give the director more flexibility during editing.

Jog The frame-by-frame movement of videotape in either a forward or backward direction for editing or other purposes. A jog is usually defined as a movement of one frame at a time, but it can actually be several frames at a time. Anything faster than this is usually called a search or shuttle.

Joystick A generic term for a device that allows the user of a videotape editing system to move the tape forward or backward at any speed from still frame to the recorder's maximum search speed. The joystick may be a stick, knob, lever, slider, or group of push buttons.

L cut *See* Split edit.

Longitudinal time code Time code that is recorded on one of two audio channels on videotape. Newer VTRs and VCRs use channel 3 as an address track to prevent cross talk and other interference with program audio. Channels 1 and 2 are used primarily for recording and reproducing program sound. On older VCRs, it is sometimes necessary to use channel 2 for time code when a channel 3 track is not available. This limits the user to one channel of program sound.

Looping The process of replacing lines of dialogue during post-production by rerecording the dialogue under controlled studio conditions. Looping is usually required if the original production dialogue quality is poor due to adverse recording conditions. *See* Automatic dialogue replacement.

Mark in/out An electronic editing term that refers to entering the beginning and end points of an edit on the fly—that is, while the VTRs are playing in real time. By pressing a single key, the Mark function enters into the computer's memory the eight-digit time code number, telling the system precisely where an edit is to begin or end.

Master The original recorded production material before it has been edited. Generally, master videotapes and audiotapes are recorded in production to be used later in editing. An edited master is then produced during post-production. The master contains the first-generation picture and sound. The term also refers to the original film-to-tape transfer.

Match cut The process of adding consecutive video or audio material to the end of an edit to extend its length or to start an optical effect transition. When an edit has been recorded on videotape and the director or editor wishes to extend the edit beyond the original out point, he or she can simply insert the extension as a continuation of the previous edit by making the out time code number of the original edit the in time code number of this new material. This eliminates the need to rerecord the entire edit plus the extension. The term *match cut* is also used to define the start of an optical effect transition such as a dissolve, wipe, fade, or digital effect.

Menu Information essential to an editing session that is displayed on the monitor of a video editing system. Some of the items that might be included in the menu are the time code location of each VTR, the number of VTRs in use, one or more lines of edit data, transition type and duration, audio/video mode type, and the title of the program being edited.

Monitor (television) A high-grade version of a television receiver constructed of industrial-quality parts. It usually does not contain any tuner or sound system found in consumer-type receivers. A monitor has video input and output connectors installed so that the signals can be fed directly into the monitor instead of going through antenna connections, as is the case with a receiver.

Monochrome A black-and-white picture containing only black, white, and varying shades of gray.

Non–drop frame time code A signal used to identify video frames (*see* Time code) developed and standardized by the SMPTE that assigns to the tape chronological time code numbers corresponding to a 24-hour clock. This type of time code has limited use in color television applications because its running time is not in agreement with clock time. *See also* Drop frame time code.

Note A typed notation generally inserted in the body of an edit list to remind the editor to perform some function at a later time. Notes may be generic or tied to a specific edit in the list. Also called a *comment.*

NTSC National Television Standards Committee. This committee was formed to recommend to the FCC a color television standard that was compatible with black-and-white television sets.

Off-line editing A term referring primarily to the decision-making or work print stage of videotape editing. The product of an off-line edit is not considered to be a broadcast-quality master but is used to generate program continuity and accurate time code data that will be used to conform a master edited tape during the on-line editing process.

On-line editing A term referring to the stage of videotape editing in which the end product is generally of broadcast quality. An on-line edit is generally used to conform and build master edited tapes from source material based on off-line edit lists.

On the fly *See* Sync roll.

Open-ended edit An edit that has a start or in point but no defined end or out point. Once started, the edit will continue until either the source material or the master tape runs out. This type of edit is useful when the exact out time is not known.

Out point The end of an edit, defined by an eight-digit time code number. Also called *out time.*

Phase, color Of or relating to color relationships or the balance of color in a video picture. A color bar standard is usually used to set the precise phase relationship of all the colors on a videotape. Also called *hue.*

Preroll The point at which all the VTRs and/or ATRs park, or pause to establish synchronization, ahead of the designated edit point prior to making an edit.

Preview An electronic editing function that allows an editor to rehearse an edit as many times as needed before recording it.

Random access The ability to retrieve information randomly from video or audio sources—that is, out of the order in which the information was originally recorded.

Real time A term defining motion occurring in a video picture at the same speed as it occurred when it was recorded. Slow or fast motion is not considered real time.

Real time editing *See* Sync roll.

Reel number A number assigned to a reel of videotape and used in the editing, reediting, or conforming process to identify each reel.

Replay A function of computer assisted editing systems that allows the editor to review an edit that was just recorded.

Ripple A term used in computer editing to refer to the process of modifying one or more edits by varying their duration or position within an edit list, thus displacing all the edits that follow by the amount of the change. Also called an *update.*

Rough cut A term used in both film and tape editing to describe the first assembly of scenes in the order of story continuity. A rough cut is almost always reedited many times to perfect continuity, pacing, and timing and to make creative changes that will result in the finished form. Also known as a *first cut.*

Search The ability of an editing system to use time code to locate a given spot on a tape quickly.

Set in/out The function of typing in an eight-digit time code number when entering data into a computer terminal. *Set in* refers to typing in an in point, and *set out* refers to typing in an out point. *See also* Mark in/out.

Slave The process of synchronizing one VTR or ATR to another device, usually by means of coordinating time code.

Slug A space left on an edited videotape that will be replaced by other material of the same length and can be used as a temporary reference. This space can be filled with video black or any other video signal. "Scene Missing" banners are one type of slug.

SMPTE Society of Motion Picture and Television Engineers. This organization was founded to promote the interests of the motion picture and television industries. Among other things, the SMPTE recommends standards for film and videotape that help reduce compatibility problems within these industries.

Splice The process of joining two edits or segments of film or videotape together either electronically or mechanically. Also, the junction of the two segments.

Split edit A transition in which the audio or video portion of the second edit leads or lags a portion of the previous edit. For example, if the sound of the incoming scene starts before the corresponding picture, it is called an audio leading video split edit. If the video of the incoming scene begins before the audio of the outgoing scene ends, it is called a video leading audio split edit. Other names for a split edit are *L cut* and *staggered edit.*

Staggered edit *See* Split edit.

Still frame *See* Freeze frame.

Submaster A high-quality videotape made from an edited master by copying from one VTR to another through time base correcting equipment. Any number of submasters can be made from a single master. A submaster is generally used as a backup or for making additional copies for broadcast, distribution, or viewing.

Sweetening The mixing, balancing, and equalizing of sound elements to create the final sound track, known as a composite sound track. Also called *dubbing* or *audio mixing*.

Sync roll A process in which edits are created by playing back multiple source VTRs at normal speed and switching the input to the recording tape from source to source, implementing either spontaneous or planned editing decisions. When this process is used on a computerized editing system, the pertinent data for each edit is automatically stored in memory as the edit is made. Also called *sync mode, sync roll* or *real time editing*.

Telecine Derived from the words *television* and *cinema,* this term refers to film-to-tape (and vice versa) transfer equipment. It is frequently used to describe either the equipment itself or the room where the equipment is housed. *See* Film chain.

Television The generic term used to describe the transmission and reception of audio and video signals in black-and-white or color.

Time base corrector (TBC) A device that corrects, within limits, the time base stability of a video signal. Many TBCs also include a video processing amplifier that allows the operator to adjust the video and color levels as well. *See* Time base stability.

Time base stability A precisely measured error signal that determines the amount of horizontal and vertical jitter, or signal fluctuation, in a video signal. The signal is found by comparing the tape playback signal with a stable reference signal.

Time code (1) An electronic signal that switches from one voltage to another, creating a series of closely spaced pulses. Each one-second portion of time code contains 2,400 pulses, or bits, which are divided equally among the 30 television frames in each second. In other words, each video frame contains 80 bits of time code data, which is used to locate information.
(2) A system of identifying material recorded on videotape by assigning each frame a chronological number based on a 24-hour clock. A time code number is composed of eight digits representing hour, minute, second, and frame. Time code allows material to be identified and located precisely for production or editing purposes. Several different types of time code are used in different situations. *See also* Drop frame time code, Longitudinal time code, Non–drop frame time code, and VITC.

Time code generator An electronic clock that generates time code and applies it to audio or videotape. Although most generators apply time code at the standard rate of 30 frames per second, some generators can produce time code at a rate of 24 or 25 frames per second. A time code generator also might be capable of generating non–drop frame or drop frame time code, as well as user bits. Also called a *time code transmitter*. *See* Time code.

Tracking A method of manually or automatically positioning the video head of a VCR or VTR so that it reads the video information precisely from the recorded tape signal, thus maintaining optimum picture quality. A mistracked video head produces video distortion in the picture.

Tracking edit A necessary but repetitive bookkeeping edit in an EDL that is used to locate play and record time code numbers, allowing a computer assisted editing system to create optical effects such as wipes, fades, and dissolves. A tracking edit is required only when generating an optical effect and is not required for simple cuts.

Transfer editing Videotape editing performed electronically by copying video signals from one VTR or VCR to another, as opposed to physically cutting or splicing segments of a videotape together. For example, a scene might be cued up on the play VTR and edited onto a master tape by electronically transferring the signal from the play VTR to the record VTR. This is the most common method of videotape editing found today.

Trim The process of adjusting an existing edit point by adding length to or subtracting length from an edit. Either the in point or the out point of an edit can be adjusted in a positive or negative direction.

Upcut An edit in which desired material is mistakenly lost due to incorrect in or out points. This can be corrected by adding material to either end of the edit and remaking it until the desired result is achieved.

Update *See* Ripple.

Vectorscope A display device used to show the electronic pattern of the color portion of the video signal. Its primary purpose is to allow the operator to adjust the color phase, hue, or chroma saturation by using a stable color reference, such as color bars, normally recorded at the beginning of each roll of videotape.

Video The electronic information that makes up the content of the television picture. This includes the black-and-white information, the color information, and the color burst signal. Although horizontal and vertical blanking signals are part of the information contained on videotape, they are not considered part of the video information because they are not seen on the screen.

Videocassette A plastic container holding prepackaged lengths of videotape that is inserted into a VCR. The most popular cassettes store tape in quarter-inch, half-inch, and three-quarter-inch widths and in 5- to 90-minute lengths.

Videocassette recorder (VCR) A unit that uses videotape enclosed in a cassette rather than on an open reel. VCRs are easy to thread, as the operator just inserts the cassette into the VCR carriage until the cassette is properly seated. The automatic threading mechanism within the VCR does the threading and unthreading with accuracy and repeatability.

Video processing amplifier (proc-amp) An output device that cleans up a video signal, removing noise and restoring the synchronizing signals to a cleaner state. A proc-amp can be used on the output of a VCR or VTR, camera, video switcher, or any other piece of equipment that might need adjustment of its output signals.

Videotape A Mylar or plastic base coated with materials capable of being magnetized in a uniform manner and of holding these patterns for an indefinite period of time. The magnetic material most often used in the manufacture of videotape is iron oxide, a very complex compound that is sprayed on the base. The iron oxide is held to the base by a binder.

Videotape recorder (VTR) An electromechanical device designed to record and play back video and audio signals on magnetic tape that is stored on open reels. The tape must be threaded carefully by hand through the machine's transport mechanism and attached to a take-up reel.

VITC Vertical interval time code. A form of time code in which the code is converted to and stored as video signal information, as opposed to being recorded on a cue channel or other audio channel. The time code is stored on lines 12 and 14 of the vertical interval, which are outside the visible picture area.

When a helical-scan VTR or VCR is slowed down below 30 frames per second (play speed), electronic circuits convert the time code data in the video signal to readable time code that is displayed on a monitor or other viewing device. When the tape is moved in either direction or is in still-frame mode, the time code displayed in the video is frame accurate and is accurately updated every field.

Waveform monitor A display device used to show the electronic pattern of the video signal, allowing the operator to adjust and maintain video signal levels.

Window dub A copy of an original videotape with the tape's corresponding eight-digit time code displayed in a rectangular area positioned somewhere in the picture area, generally at the bottom of the screen. This window area can be surrounded by a black box so that the time code numbers will stand out against a light-colored background. The window dub can be used only as a viewing copy or work tape, since the numbers cannot be removed from the picture.

Wipe A transition between two or more video sources that is seen as a moving line, circle, or other complex pattern with a soft, hard, or bordered edge separating the video sources. *See* Duration.

Selected Bibliography

Anderson, Gary. *Video Editing and Post-Production. A Professional Guide.* White Plains, NY: Knowledge Industry Publications, Inc., 1984.

Anderson, Gary. *Electronic Post-Production: The Film-to-Video Guide.* White Plains, NY: Knowledge Industry Publications, Inc., 1986.

Browne, Steve. *Videotape Editing: A Post-Production Primer.* Boston: Focal Press, 1989.

Dmytryk, Edward. *On Film Editing.* Boston: Focal Press, 1984.

EECO, Inc. *The Time Code Book.* 4th ed. Santa Ana, CA: 1987.

Millerson, Gerald. *The Technique of Television Production.* 11th ed. Boston: Focal Press, 1985.

Millerson, Gerald. *Video Production Handbook.* Boston: Focal Press, 1986.

Reisz, Karel, and Gavin Millar. *The Technique of Film Editing.* 2d ed. Boston: Focal Press, 1968.

Robinson, J.F., and Stephen Lowe. *Videotape Recording.* 3d ed. Boston: Focal Press, 1981.

Shetter, Michael. *Videotape Editing: Communicating with Pictures and Sound.* Elk Grove Village, IL: Swiderski Electronics, 1982.

Weise, Marcus. *Videotape Operations.* Woodland Hills, CA: Weynand Associates, 1984.

Weynand, Diana. *Computerized Videotape Editing.* Woodland Hills, CA: Weynand Associates, 1983.

Weynand, Diana. *The Post Production Process.* Woodland Hills, CA: Weynand Associates, 1985.

Wiegand, Ingrid. *Professional Video Production.* White Plains, NY: Knowledge Industry Publications, Inc., 1985.

Index